COASTAL NAVIGATION

Jeff Toghill

W·W·NORTON & COMPANY
New York   London

Published simultaneously in Canada by Penguin Books Canada Ltd.,
2801 John Street, Markham, Ontario L3R 1B4.

Printed in the United States of America.

ISBN 0-393-30293-8

W. W. Norton & Company, Inc., 500 Fifth Avenue, New York, N.Y. 10110
W. W. Norton & Company Ltd., 37 Great Russell Street, London WC1B 3NU

1 2 3 4 5 6 7 8 9 0

Contents

Introduction

Most books on navigation are based on the subject as practiced aboard ocean-going freighters or naval vessels. Yet navigation on a yacht is as far removed from navigation on a big ship as a well-aged wine is from grape juice. Few yachts carry the sophisticated equipment fitted aboard naval or commercial vessels, and few yachtsmen or women have the time or the inclination to study navigation in depth as do full-time professional navigators.

Yet navigation is not a subject to be taken lightly since the safety of the vessel, however large or small, is at stake. This book streamlines and simplifies 'big ship' navigation to a practical level for small-boat operators without reducing the accuracy of the workings. It provides a basic tutor for yacht navigators planning to make coastal passages which will, for the most part, keep them within sight of land.

In short, it provides offshore yachtsmen or women with all they will require to get from arrival and departure points along a coastline with the minimum of fuss but with maximum safety. It is based on my 35 years navigating yachts in all corners of the world and while the emphasis throughout is on simplicity, in no way is safety sacrificed.

Jeff Toghill
1986

Defense Mapping Agency Hydrographic/Topographic Center charts and some illustrative pages from the Center's publications reproduced by courtesy of the Defense Mapping Agency Hydrographic/Topographic Center, Washington DC, USA.

Chapter 1   The Chart

What is a chart?

The whole of coastal navigation is based upon the nautical chart. The chart is, in effect, the mariner's road map. But while there are many similarities between the familiar road map and the nautical chart, there is one very big difference between them:

> The road map deals solely with the land and virtually ignores anything seaward of the coastline.

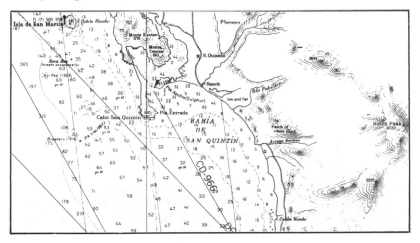

The nautical chart deals solely with the coastline and the sea and ignores most things inshore of the coastline.

These are very broad descriptions, of course, and there are exceptions in both cases. The road map, for instance, will carry details of islands off the coast if there is anything of interest about them. Similarly, the nautical chart will carry details of mountains, towers or other high objects which may be located well inland from the coast but which are visible from sea, and of use to the marine navigator.

7

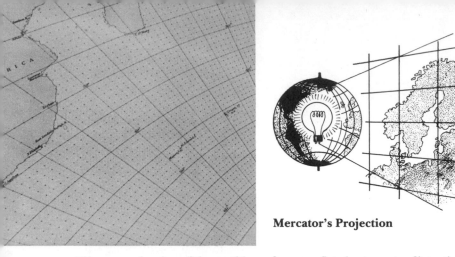

Mercator's Projection

True reproduction of the earth's surface on a flat chart creates distortions of the land areas.

How a chart is made

Again like a road map, the nautical chart is the representation on a flat sheet of paper of a section of the earth's surface. Because the earth is round, this requires a special construction process to avoid distortion. Imagine part of a tennis ball being flattened. The edges would split and the surface would become distorted.

While it is impossible to flatten all sections of the earth's surface into chart form without such distortions, small areas, particularly if they are not too near the poles, can be successfully reproduced on a flat surface by a method known as Mercator's projection. All modern nautical charts used for coastal navigation are made by this projection, and, providing they cover only relatively small areas, there is insufficient distortion to worry about.

Probably the best way to illustrate the projection of the earth's surface onto a flat map is to imagine the globe being transparent, with the land masses marked on it. If a light is placed inside this transparent globe and the whole unit is placed near a wall, the land masses will be projected as shadows onto the flat wall.

Obviously, much of the projection will be distorted and the distortion will be greatest towards the extremities of the wall, so small sections at the very center of the projection are used to make charts and the distortion is then at a minimum. The globe can be rotated so that each area can be located in the center and a chart taken off. While this is not the actual technique used in making nautical charts in practice, it provides a good illustration of the principle involved.

8

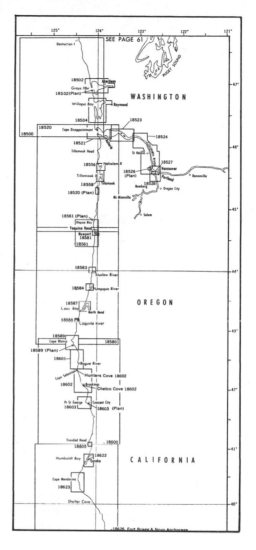

Extract from a typical chart catalog.

Where charts are obtained

In most countries nautical charts are made by the hydrographic service of the navy. In the United Stated of America nautical charts are obtainable through the National Ocean Survey or the Defense Mapping Agency, Hydrographic/Topographic Center in Washington D.C. In many ports, particularly where there are no official establishments, chart agents are appointed. These may be bookshops or suppliers of nautical equipment, and carry supplies of nautical charts and marine publications.

9

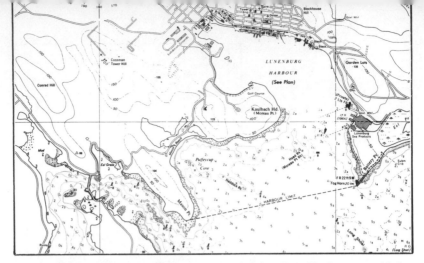

The medium scale chart offers a great deal of detail and is the most widely used chart. Large scale chart of the port is indicated for making an entrance.

A chart catalog, or index, with details of all published charts, enables the navigator to determine at a glance exactly which chart he requires. Special charts for yachtsmen are also available in many countries.

The scale of a chart

Since there is a nautical chart for every ocean, every coastline and every port in the world, it follows that there must be charts of differing scales.

On a 'Sailing' chart, for example, there would be little need for great detail since, for the most part, the area covers vast open spaces. On a 'Coast' chart, by contrast, fine detail of every inch of the coastline is required so that vessels can navigate safely through the many shoals and hazards.

There are three major scales for nautical charts, and each covers a specific area of navigation:

1 Small scale charts cover large stretches of ocean and carry only limited detail of coastlines. These are known as *Sailing charts*.

2 Medium scale charts, which carry sufficient detail along varying stretches of coastline to enable a boat to be successfully navigated offshore. These are *Coast charts*.

3 Large scale charts, which carry fine detail of virtually every inch of estuaries, ports and rivers, to enable boats unfamiliar with the area to make a safe entry. These are termed *Harbor charts*.

10

Apart from a full range of safety equipment, the well-equipped coasting yacht will carry a complete portfolio of inshore charts

Small-craft charts are also available from the National Ocean Survey. These are designed primarily for boatowners using inland waterways and are a useful aid to boats navigating in such water.

The would-be navigator, setting off on a passage, can select from the chart catalog all the charts he will require to cover all aspects of navigation and safety *en route* without cluttering up his boat with unnecessary folios.

If he is making an ocean crossing, for example, he will select a Sailing chart to cover the ocean area, two or three Coast charts to cover his landfall and any coastal passages involved, and Harbor charts of his departure and arrival ports or any other places he may visit either for interest or for shelter.

Metric charts

A gradual conversion to metric charts is taking place in many parts of the world. It is currently possible to purchase both fathom and metric charts, sometimes of the same area, often of the same coastline. Needless to say it is vitally important to the navigator to know which chart he is using. This information is carried in the chart title and should be checked when the chart is purchased.

NOVA SCOTIA - SOUTHEAST COAST

LUNENBURG BAY

Surveyed by F.L. DeGrasse and assistants, 1959-62

Cross Island (Geod.) △ : Lat. 44° 18′ 46″18 N., Long. 64° 10′ 12″58 W.

Bearings refer to the True Compass and are given from Seaward (thus 295° etc.)

SOUNDINGS IN FATHOMS
(under 11 in fathoms and feet)
reduced to Lowest Normal Tides

Water areas with depths of 6 fathoms and less are tinted blue except in dredged areas
Underlined figures on drying banks or in brackets against drying rocks
express heights in feet above the datum of soundings; all other heights
are expressed in feet above Higher High Water, Large Tides

For Symbols and Abbreviations, see Chart No. 1

Natural Scale 1 : 18,000

Projection : Polyconic

TIDAL INFORMATION

PLACE	Height above Datum of Soundings				
	Large Tides		Average Tides		Mean Sea Level
	Higher H.W.	Lower L.W.	Higher H.W.	Lower L.W.	
	feet	feet	feet	feet	feet
Lunenburg	7·6	0·8	6·5	1·9	4·2

BENCH MARK
The datum of this chart at Lunenburg is 12·57 feet below a H.S. bronze tablet stamped "HS-2 1960" set in a concrete pillar at the corner of the Acadia Supplies warehouse.

LONGITUDE

Chart title *(above)*. Global grid of Latitude and Longitude *(top right)*. The grid as seen on a chart *(bottom right)*.

The chart title

Every chart carries a title with a wealth of information of use to the navigator. Apart from the obvious description of the area covered by the chart, details relating to the scale, the soundings, tidal anomalies, military areas, special areas, specific dangers and any other factors which will assist in safe navigation are listed in the chart title.

Latitude and longitude

Mercator charts are covered by a 'grid' pattern indicating lines of latitude and longitude which match those on the earth's globe. The lines which run vertically up and down the chart are meridians of longitude, and those which run horizontally across it are parallels of latitude. Their use will be seen in a later chapter when the chart is used for plotting.

The scale across the top and bottom of the chart is the longitude scale and is marked with degrees and minutes of arc (60 minutes equals 1 degree). Longitude commences at 0° on the Greenwich meridian (London), and runs east and west 180° to the opposite side of the world. Thus the graduations on the top and bottom of the chart will be in degrees and minutes of longitude east or west of Greenwich and marked accordingly.

On either side of the chart are the graduations for the latitude scale. Latitude commences on the equator and is measured north and south to the poles. The graduations are the same (degrees and minutes) as longitude, but the two are dissimilar in scale and must never be confused.

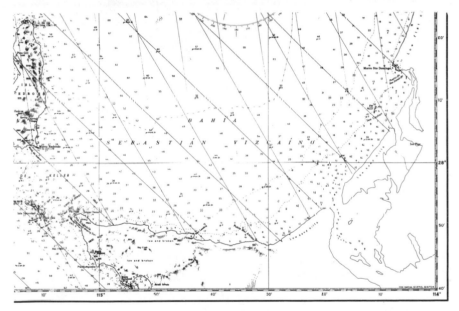

Distance scale

Distances at sea are always measured in nautical miles, and thus the distance scale is the same on every chart regardless of whether it is graduated in meters or fathoms. Any distance scale which may be incorporated in the title—or anywhere on the chart for that matter—can be ignored, since it is related to land miles. The latitude scale on either side of the chart (*never* the longitude scale at top or bottom) is used to measure distance, thus:

One minute of latitude =One nautical mile.

13

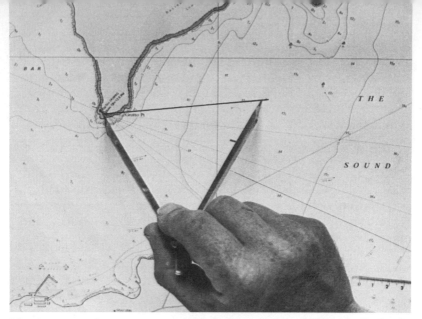

Distances on a chart are measured by dividers and read off on the Latitude scale

Soundings

The depth of water indicated by figures all across the seaward area of the chart may be in fathoms or meters (1 fathom=6 feet). Close inshore these depths may be represented in smaller denominations, either as fathoms and feet or as decimals of a meter. This is probably the most confusing and dangerous aspect in the conversion to metric charts and the navigator must be very aware of it.

The depths are indicated by a normal sized sounding figure with a smaller figure below and to the right of it. They are read thus:

Fathoms: 5_2 (5 Fathoms 2 feet=32 feet)
Meters: 5_2 (5.2 meters=17 feet approx)

The danger of confusing the two systems is obvious from this example.

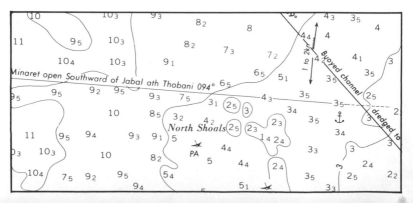

Chart datum represents the structure of the waterway at lowest low tide. Tide heights must be added to chart figures.

The charted depths, known as soundings, are reduced to the lowest mean low tides experienced in the area. Thus it is safe to say that, *with only rare exceptions*, the soundings on the chart indicate the least water over the sea bottom that will be experienced in a normal tidal cycle. Or, to put it more practically, there will rarely be less water than that indicated by the soundings. The level to which soundings are reduced is usually Mean Low Water Springs (MLWS) or Lowest Astronomical Tide (LAT).

Chart datum

This is the term given to the level of soundings on the chart. It is the basis on which all tide tables are calculated. Thus, to find the depth of water over the sea bottom at any time, the height of the tide, from the tide tables, is added to the sounding on the chart. This is described in more detail in a later chapter dealing with tides and tidal streams.

Colored shading

Modern charts employ a system of color shading or tinting to give greater emphasis to certain depths. Particularly is this the case with inshore soundings. Shallow waters may be indicated by a deep blue color, (blue-gray if they will dry at low tide), which

15

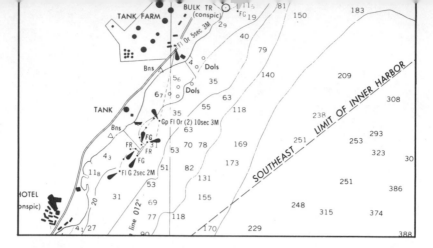

fades out to a paler blue as the water becomes deeper. The land is usually shaded gray although some modern metric charts may use yellow or a similar contrasting color, and certain danger areas, or areas of military or commercial use, may be covered by a purple tint.

Navigation lights, which are the navigator's signposts, as it were, are indicated on the chart by a purple 'flare' symbol. While there are other shore lights that can be used in navigation, those indicated by the purple flare are the only completely reliable navigation beacons.

Contour lines

Although mountains are not used widely for navigation unless they are very prominent or have some particularly obvious characteristic, their presence may be indicated by the contour lines used on normal land survey maps.

Offshore, the undulations of the sea bed are indicated by contour lines often similar to those of the mountains. Full details of both systems are given on Chart No.1, described later in this chapter.

Careful chart reading avoids hazardous situations such as this

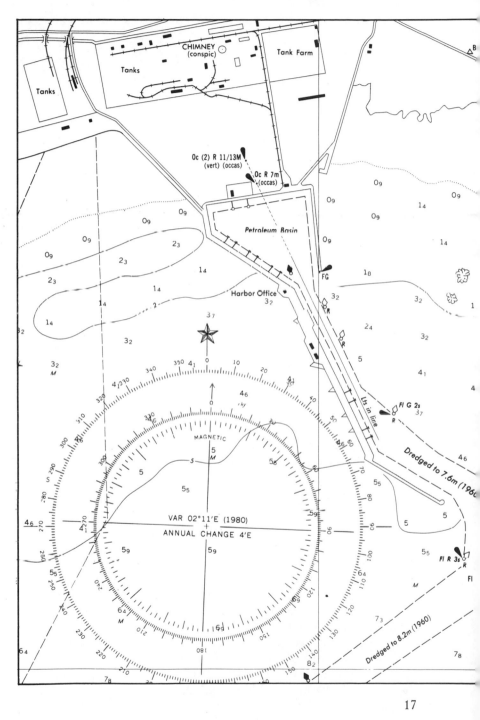

CHIMNEY
(conspic)

Tank Farm

Tanks

Tanks

Oc (2) R 11/13M
(vert) (occas)

Oc R 7m
(occas)

Petroleum Basin

0_9

0_9

0_9

1_4

0_9

0_9

1_4

0_9

2_3

0_9

0_9

2_3

1_4

FG

1_8

2_3

1_4

Harbor Office

3_2

3_2

3_2

1

1_4

1_4

2

3_7

2_4

3_2

R

3_2
M

R

5

4_1

4

340 350 4_1 0 10 20

330

320

46

Fl G 2s
3_7

R

310

46

300

5_9

MAGNETIC

Dredged to 7.6m (196

290

5_5

5

S

280

5_5

5

5_9

270

4_1

4_6

5

260

5_9

VAR 02°11'E (1980)
+
ANNUAL CHANGE 4'E

5_9

5_5

Fl R 3s
R

250

5_5

Fl

6_4

7_3

M

230

7_8

Dredged to 8.2m (1960)

6_4

8_2

7_8

7_8

Lts in line

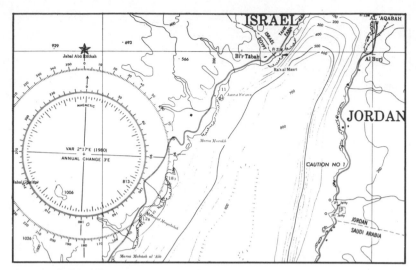

A remarkable illustration of sounding contours is shown on the chart of the Gulf of Akaba

Towns, villages etc

Since streets, houses, shops and the general makeup of towns and villages are of no interest to the navigator, they are rarely marked on the chart except in basic form. Particularly is this the case when they are inshore from the coastline. However, any prominent object which can be seen from seaward and used for navigation is pinpointed and marked very distinctly.

Typical of the sort of objects which fall into this category are church steeples, factory chimneys, water towers, radio towers, etc. Indeed, any object anywhere which can be seen clearly from seaward will be marked on the chart.

Details on a chart

One of the biggest problems in constructing a nautical chart is to include all the detail necessary to enable it to be used for safe navigation. The motorist has road signs all along his route to supplement his road map. But there are no signs to assist the boat navigator and all his information must be carried on the chart. To avoid completely cluttering the chart, the wealth of detail relating to both sea and coastline is abbreviated or indicated by symbols.

A special chart containing the key to all symbols and abbreviations used on nautical charts is published jointly by the Defense Mapping Agency Hydrographic/Topographic Center and the Department of Commerce, National Ocean Survey. It is approp-

O. Dangers

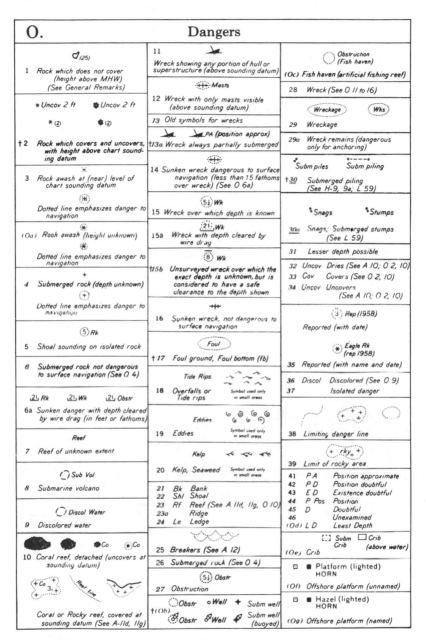

A page from US Chart No.1

riately called Chart No. 1 and should be kept aboard every boat for easy reference. The conscientious navigator will see to it that he knows by heart the more common symbols, particularly those used to indicate dangerous areas.

Symbols and abbreviations

Abbreviations used on a chart are usually in the form of a single capital letter. The letter M, for example, when printed on an open stretch of water, indicates the nature of the sea bottom—mud. If one letter is not sufficient, then two or perhaps three may be used, thus: Tr—tower; Whf—wharf; Bn—beacon.

While abbreviations are often fairly obvious, the same cannot always be said of symbols. A dangerous rock, for example, is illustrated by a small cross, a lighthouse by a tiny stellate. These symbols are all listed in detail in Chart No. 1, but since there is always the possibility that a copy of this may not be kept on board, or that there is not time to go below and read it up when in narrow or dangerous waters, the wise navigator, as mentioned, will know by heart the more common symbols so that he can identify them quickly on the chart.

The following illustrations are of such symbols and the objects they represent.

21

Radar prominent areas

Radar, while still limited in very small boats, has given the navigator a new string to his bow. Electronic navigation has come into its own in recent years and charts have been kept in line with this development. One of the weaknesses of radar is its tendency to show up different types of objects with different emphasis on the screen. A high cliff will show up clearly while a low scrubby shore-line may not show up at all.

Some charts have been specially drawn to help overcome this problem by emphasising areas which are 'radar prominent'.

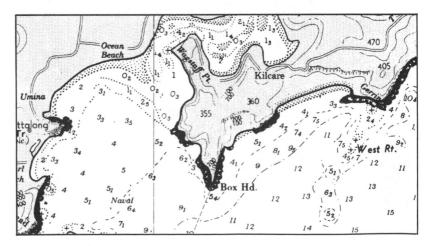

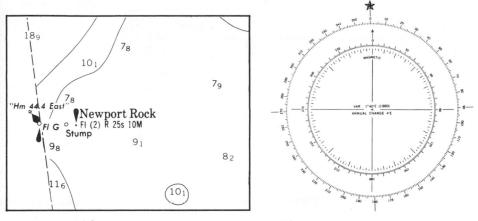

A special feature.   **The compass rose.**

These areas of coastline are printed more heavily on the chart and thus stand out against the rest of the coastline, much as they would when viewed on the radar screen.

Special features

Outside the normal symbols and abbreviations there may be need to convey some special information about an object marked on the chart. In this case a more extensive abbreviation than a single letter may be used. Most of these are self-evident.

An object which stands out well from its surrounding environment, for example, will be marked with the abbreviation (*conspic*). An area in which a light cannot be seen will be marked (*obscd*), indicating that the light is obscured in this zone. Similar abbreviations may be used for a whole variety of objects where there is some special navigational feature about them.

The compass rose

Since most navigation on a nautical chart involves the use of the compass, a reproduction of a compass card is printed at strategic points across the face of every chart. These are termed compass roses and their positioning is such that there is always one close at hand no matter where on the chart the navigator is working. In addition to the normal compass graduations each compass rose contains details of the magnetic variation in that area.

The use of the compass rose for chartwork and the information related to variation are contained in Chapter 3 of this book.

23

No. 26

1985

NOTICE TO MARINERS

PUBLISHED WEEKLY BY THE
DEFENSE MAPPING AGENCY HYDROGRAPHIC/TOPOGRAPHIC CENTER

PREPARED JOINTLY WITH THE
NATIONAL OCEAN SERVICE AND U.S. COAST GUARD

CONTENTS

PAGE

29 JUNE

Typical Notice to Mariners.

Correcting the chart

Like all maps, charts can quickly become outdated by changes to the coastline or to navigational objects. Removal of a prominent shore object such as a radio tower or chimney, for example, or changes in the light shown from a lighthouse due to maintenance work are but two factors which could cause confusion to the offshore navigator.

Such changes are published weekly in the Defense Mapping Agency's *Notice to Mariners*. From the details in these notices, the navigator can correct his charts and keep them up to date with events which may affect his navigation. Permanent changes are usually made on the chart in purple ink, temporary changes in pencil.

Details of every facet of a river or estuary are among the items contained in the *Sailing Directions*.

Whenever a change is made on the chart, the date and the number of the *Notice to Mariners* used is entered in the lower left hand corner of the chart. In this way, a glance at the listed numbers will indicate the extent to which the chart has been corrected. The onus of keeping a chart up to date with the latest *Notice to Mariners* lies with the navigator. A Summary of Corrections is issued periodically by the Defense Mapping Authority Hydrographic/Topographic Center. *Local Notice to Mariners* are published by the U.S. Coast Guard to supplement the D M A H T C weekly publication.

Sailing Directions

Sailing Directions, also called 'Pilots' are a series of volumes designed as a guide for navigators sailing in unfamiliar waters. They cover mostly foreign ports and coastlines and the ocean regions of the world. The United States coastline is covered by a similar series of publications called *Coast Pilots*, and *Great Lakes Pilots* covering the Great Lakes, issued by the National Ocean Survey. These are updated annually and provide a valuable supplement to the chart for navigators sailing along an unfamiliar part of the coastline or entering a strange port.

Both *Sailing Directions* and *Coast Pilots* are corrected and updated from information contained in the *Notice to Mariners*.

25

Chapter 2 Navigation Equipment

Before leaving harbor to make a coastal passage, the navigator has many things to do. Firstly he must plan the passage to ensure that the boat makes the shortest trip consistent with safety. Then he must ensure that she is ready for the trip in terms of navigational equipment and charts. He must satisfy himself in regard to currents and tides, weather and other factors which may affect the boat during the course of her passage.

Perhaps the first thing which can be done well in advance is to ensure that the boat is fitted with the correct navigational gear to enable the navigator not only to lay off her courses, but also to plot her track and keep check of her progress as the passage proceeds.

There is a lot of preparation work behind making a coastal passage.

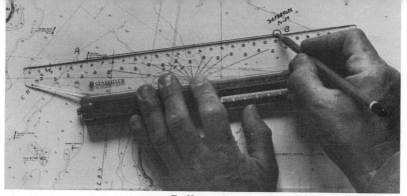

Roller rules.

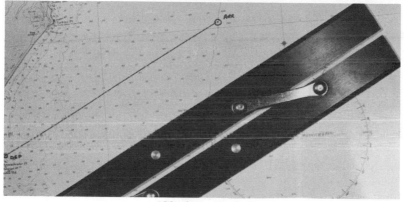

'Clackety-clacks'!

Chart instruments

Parallel rules

These are double-edged rules which are used to transfer courses, position lines and bearings across the chart. They are essential for coastal navigation and can be obtained in two principal forms— sliding rules which shuffle across the surface of the chart, or roller rules which are, as their name denotes, fitted with small rollers to enable them to be rolled parallel across the chart.

For small craft the roller type is probably more suitable since the sliding type tends to stick on any wet spot on the chart, and there would be few yachts which do not get water on the chart at some stage in a coastal passage.

By placing one edge of the parallel rules against a line (or a compass bearing), the line can be transferred to any part of the chart simply by moving the rules gently across the surface. Patent types of rules for laying off courses are also available and these can be quite useful if one becomes adjusted to using them.

27

Protractors are also popular with some yacht navigators who find them easier to use than parallel rules.

Dividers

These are used for measuring distances, as described in Chapter 1 and are as essential as the parellel rules. The 'one-handed' variety illustrated are the easiest to use and provide a good span to cover long distances across the chart.

A pair of compasses

Not to be confused with the marine or magnetic compasses, these are simply the compasses from the school geometry box used for drawing circles. They can be useful during the laying off of courses, when planning a passage, as described in Chapter 5.

Pencils and rubber

These are pretty obvious requirements, but it is important to note that the pencil should be of soft grade. Charts are made of fairly tough paper to facilitate easy rubbing off when plotting is finished, but if a hard pencil is used, an imprint may be left on the surface of the chart which will be hard to eradicate.

Stop watch

Always useful when timing the cycle of lights and beacons.

Safe arrival on a passage is assured if the boat carries the necessary equipment both above and below decks.

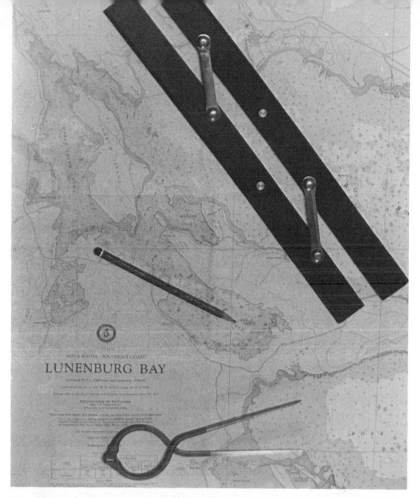

The simple requirements for chartwork.

Chart table

The size and fitting of a chart table will depend on the size and roominess of the boat. Some navigators work on the saloon table, others have a specially fitted navigation area with an established chart table. But for the average small yacht, a flat surface roughly the dimensions of half a chart is all that is required and this can be fixed temporarily over a bunk or some similar spot.

The half-chart size enables the chart to be folded just once thus avoiding too many creases on its surface while permitting reasonable working room. An underlay of flat rubber sheeting between the chart and the table prevents the paper slipping or the table getting marked by pricks from the dividers. A 'fiddle' or raised

29

Simple chart table for a small yacht.

edge around the table prevents everything sliding to the deck whenever the boat heels.

Electronic instruments

Depth or echo sounder

At one time, the depth of the water was found by use of a 'lead line'—a marked line with a weight at one end. The weighted end was dropped into the water until it touched bottom and the depth measured off the line where it broke the surface. A few lead lines are still in use, but for the most part boats nowadays use the electronic depth sounder for finding the depth of the water under the keel.

This sounder works on the principle of an electronic sound pulse which is transmitted from a transmitter on the bottom of the boat. When this pulse hits the sea bed, it 'echoes' back and is picked up by a receiver, also in the bottom of the boat. The instrument then measures the time taken for the pulse to echo and, knowing the speed at which sound travels, indicates the depth of the water on a scale.

There are many types of readout on echo sounders. Some have

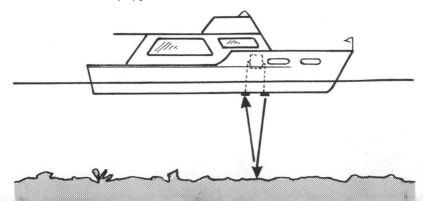

Above: A typical depth sounder with digital readout.
Right: A small boat RDF unit.

a dial around which a neon 'blip' travels to indicate depths beneath the hull, while others have a continuous roll of graph paper, rather like a barograph, on which a stylus draws a trace of the bottom of the sea as the boat passes over it. A popular modern system is a simple digital or LED readout. Most of these systems are accurate, although the trace method is preferred by most professional navigators.

A recent development of the echo sounder uses the submarine 'sonar' system which enables the pulse to be directed at angles all round the boat, thus picking up objects other than those directly beneath the hull.

Radio direction finder

Known as RDF, this is an electronic means of taking a directional bearing of a shore transmitter. The transmitter sends out a continuous coded signal, the identity and frequency of which can be found in the publication *Radio Navigational Aids*. On board the boat, this signal is picked up by a special receiver with a tuneable antenna. As the antenna is turned, the signal fades away until it dies out altogether then, as the antenna continues to turn, it fades up again.

The direction of the station is found when the signal cuts out totally at a point known as the 'null', and by means of a nearby compass, a bearing can be taken and plotted on the chart (see Chapter 6).

Much the same effect can be obtained with a simple transistor radio tuned in to a commercial broadcast station. The 'loop' antenna inside the transistor causes the audio signal to increase and fade as the set is turned towards and away from the station.

Special RDF transmitters are set up along the coastline and often linked together so that the navigator can obtain a series of bearings of different stations and thus plot his position. Automatic RDF receivers (ADF) make the work of radio navigation much easier.

The log

As with the depth sounder, the electronic log is a modern addition to small boat equipment. Previously the distance a boat travelled through the water was recorded by a spinner on the end of a line dragged behind the boat. As the spinner turned, it twisted the line which, in turn, recorded the 'twisted' distance on a dial at the stern of the boat. This type of log, known as the 'Taffrail Log', is still widely used today and is surprisingly accurate.

Most electronic logs depend on a small impeller which is attached to the hull of the boat. As the hull moves through the water the impeller turns and the distance is recorded on a dial somewhere in the boat. These logs may sometimes be driven mechanically by means of a twisting cable, but most are connected electronically. The dial records the distance travelled and sometimes also the speed of the boat.

Radar

One of the most useful of all electronic navigation instruments radar was for many years restricted to large craft. It was too big, too bulky and far too expensive for all but luxury yachts, but

A purple circle is the chart symbol for an RDF station.   **The Taffrail or towed log.**

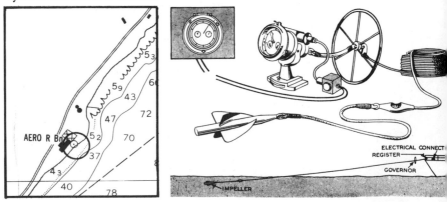

ELECTRICAL CONNECT
REGISTER
GOVERNOR
IMPELLER

AERO R Br

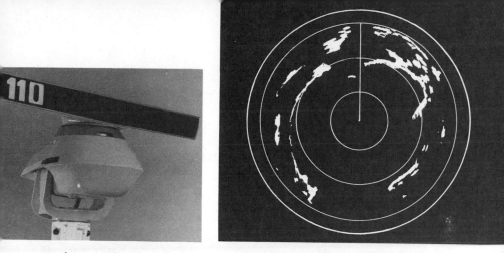

transistor technology has reduced its bulk, and mass production has brought down the price to acceptable levels so that nowadays the average yacht or motor cruiser can fit a small radar set.

Radar works on much the same principle as the depth sounder in that it transmits an electronic pulse and times the echo to determine distance. The difference in practical use, however, is that the radar set transmits its pulse in a 360° circle. The echo from surrounding objects is not recorded as a figure on a dial or a trace, but is laid out in the form of a map of the surrounding area with the boat in the center. This type of display is known as the 'Plan Position Indicator' (PPI).

Obviously, this makes life very easy for the navigator, for a glance at the radar screen can tell him just where the boat is positioned in relation to the surrounding coastline, and he can compare this with an identical chart. If he wishes, he can also take off bearings and distances of prominent objects in order to plot the boat's position accurately.

As mentioned in Chapter 1 however, radar has one major weakness in that it tends to echo badly off certain terrain. Thus the picture on the screen is not always an accurate one and does not compare easily with the chart. The PPI display must be carefully interpreted so that mistakes are not made. Modern charts are sometimes marked to indicate which areas offer best radar 'echoes'.

Loran C

Covering most of the coastline of the U.S. and adjacent waters, Loran C is a radio navigation aid which picks up transmitted radio signals from which it determines the boat's position. It is simple to operate and provides an accurate latitude and longitude readout.

33

Chapter 3   The Marine Compass

The gyro compass

This is without doubt the best type of compass for marine work and is fitted to most commercial and naval vessels. Unfortunately it is not easily adapted to small boat use, mainly because of its complex makeup. The compass comprises a free-spinning gyroscope wheel which aligns itself in the true north/south line. Its advantages lie in its freedom from the errors which plague the magnetic compass, but its drawbacks lie in the difficulty of fitting and maintaining such complex equipment in small craft. For this reason, only the magnetic compass will be dealt with in this book.

The magnetic compass

In its most basic form, the magnetic compass can be considered to be the needle-type 'Boy Scout' compass familiar in so many different sports. The needle, which is itself magnetised, lies in the magnetic field of the earth in such a way that it points towards the magnetic north pole no matter how the compass bowl is turned.

The marine compass

The principle difference between a needle compass and the marine compass is that in the latter the swinging needle is replaced by a swinging card. This is purely to allow easier reading of the compass when the boat is moving around in a seaway, for basically both compasses are the same. Instead of using a needle swinging across a card, the marine compass has a free swinging card to which magnetic needles are attached with their north pointing end in line with the north point on the compass card. When the needles swing to take up their magnetic north/south line, they swing the card with them.

Boat compasses
come in a variety of
shapes and sizes.

An ideal small yacht
compass.

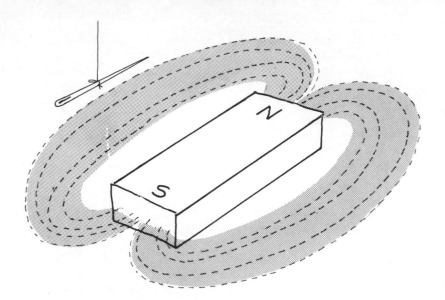

Earth's magnetic field

To understand how a marine compass is used it is necessary to understand the basic magnetic field of an ordinary magnet. An exercise used to demonstrate magnetic fields to school students provides a good illustration. A magnet is placed beneath a sheet of paper covered with iron filings, the filings form themselves into a pattern illustrating the lines of magnetic force radiating from one pole to the other. The needle of a compass placed within this field will align itself with the lines of force that surround it.

Now think of the world as a magnet. The lines of force radiate from the poles in just the same way and the needle of a compass placed within the earth's field will align itself with these lines of magnetic force. Since they run between the north and south poles, the needle will align itself in the north/south direction. Unfortunately the magnetic poles are not situated in exactly the same place as the true poles and so an error occurs in the reading of the compass.

Variation

This is the name given to the error caused by the difference in position between the true and magnetic poles. As its name denotes, it varies from place to place across the world, but it is accurately tabulated for the navigator's use. It can be described as follows:

Variation is the error in the compass caused by the earth's magnetism. It is always named E or W according to which direction the compass needle is deflected away from true north.

Deviation

The second of the two errors which affect the magnetic compass, deviation is caused by the magnetic influence of anything near the compass. Sometimes placing a metal knife alongside the binnacle, for example, will cause a deflection of the compass needle and result in a deviation error. Steel in the construction of the boat's hull, electric circuits, motors, and so on, can all affect the compass and create a deviation error. It would be safe to describe the error as follows:

Deviation is the error in the compass caused by the boat's magnetism. It is always named E or W according to the direction the compass needle is deflected from true north.

To find variation error

This is easy. On every chart there are a number of compass roses as described in Chapter 1. In the center of each compass rose is listed the variation for that area and the amount it is likely to change in one year, which is usually fairly small.

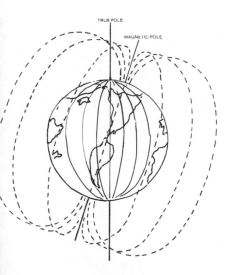

Earth's magnetic field.

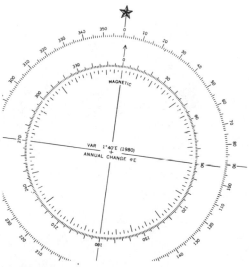

Variation is written across the compass rose.

37

To find deviation error

This is not so easy, for deviation can be caused by not one, but a combination of many factors. To begin with, every new fitting, new stores, or new equipment placed on the boat can change the error of deviation. Providing the new acquisitions are not too magnetic and they are kept at least 1 metre away from the compass binnacle, they should not have too much effect, and this is worth remembering when fitting out a boat

    Deviation can change with each change in the boat's direction, which creates another problem. Finally, the boat itself, particularly if she is of steel construction, will have become a magnet in her own right during building and, as can well be imangined, this will also play havoc with the accuracy of the compass in the binnacle.

    The best way to find the deviation error is to engage a professional compass adjuster and have him calculate and then eliminate the error, or if it cannot be eliminated, then minimise it and tabulate the residual error on what is known as a deviation card.

Large motor cruisers and yachts with steel hulls are most susceptible to deviation problems.

DEVIATION CARD

VESSEL *Cygnet*   TYPE *24' Sloop*

BOAT'S HEAD		DEVIATION
N	000°	0°
NE	045°	1° East
E	090°	2° East
SE	135°	1° East
S	180°	0°
SW	225°	1° West
W	270°	2° West
NW	315°	1° West
N	360°	0°

The deviation card

When a boat is checked for deviation she must be checked on all headings since, as mentioned, deviation varies according to the course being steered. The compass adjuster will swing the boat through the major compass points and determine the deviation on each point. He will then list the deviation error on each heading on a deviation card, a sample of which is given here. Thus the navigator can, by referring to this card, determine the deviation error on whatever course he is planning to steer.

39

Care must be taken with compasses mounted among electrical instruments or local deviation may occur

The compass card

For many years the traditional mariner's compass carried a card on which were ornately printed the cardinal points of the compass as well as three-figure notation. Modern compasses, however, have mostly done away with the cardinal system and have only three-figure notation (0°-360°) inscribed on the edge of the card.

The card commences at 0° (due north) and travels in one degree units through 90° (due east), 180° (south), 270° (west) back to 360° or 0° at due north.

The lubber line

The lubber line is the term given to the mark on the bowl of the compass which represents the centerline of the boat. Thus, when steering a course, the card is swung until the figure representing the required course is aligned with the lubber line.

The master compass

This is the compass located near the helm by which all courses are steered. It is usually the largest and most accurate compass in the boat and is mounted in a binnacle with screening and lighting to make reading easier at night.

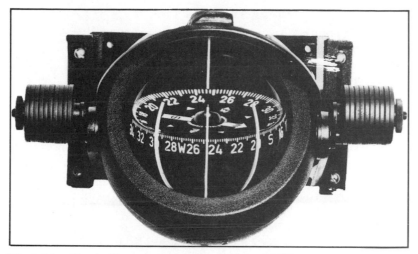

The lubber line indicates a compass reading of 247°.

The master compass is always located near the helm.

The repeater or tell-tale compass

Any small compass situated in the boat—on the flybridge of a motor-yacht, above the skipper's bunk, or on the navigator's table—is referred to as a tell-tale compass. It may also be referred to as a repeater compass.

The hand-bearing compass

The most useful compass on small boats other than the master compass is the hand-bearing compass. It is completely portable and can be used in a variety of places around the boat. As its name denotes, it is principally used for obtaining bearings and since, particularly on a heeling yacht, it is not always possible to obtain bearings from the cockpit, it may be carried to any strategic position where good bearings can be taken.

Usually this compass is fairly small and light and has a handle for convenience when holding. Since it is used mostly at eye level, it is often fitted with a prism or mirror to reflect the figures of the compass card.

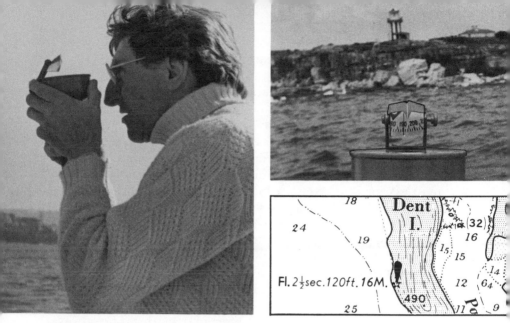

A. Sighting up the object in the compass.
B. Lining up for a reading.
C. Indentifying the object on the chart.

Taking a compass bearing

A shore object, of which a bearing is to be taken, is first identified on the chart. The hand-bearing compass is then held up to eye level until this object appears in the compass sight. The lubber line and the figures on the relevant section of the compass card are reflected in the prism or mirror directly below this sight. When the object, in the sight, the lubber line and the compass figures are all in alignment, the figure against the lubber line is read off as the compass bearing of the object from the boat.

It follows that if the shore object lies along the compass bearing from the boat, then the boat must lie along the reciprocal or reverse bearing from the object. This is one of the most basic factors employed in finding the boat's position by compass bearings.

Good accurate bearings can usually be obtained from objects such as:

A headland terminating in a steep cliff.
A small island (peak or edge).
TV towers
Lighthouses

43

Conspicuous factory stacks
Oil rigs
Triangulation (survey) stations

Liquid damped compasses

Because of the violent motion of a small boat in a seaway, the compass card is liable to charge around on its pivot making it almost impossible for the helmsman to follow. For this reason most marine compasses are filled with liquid and fitted with a 'damping' device which slows the swing of the card and absorbs the violence of the boat's movement, thus making steering and navigation easier and more accurate.

Compass error

As described earlier in this chapter, every boat is affected by variation and deviation. A good compass adjuster will eliminate the deviation if it is small, but, unfortunately, this is not always possible and often both errors are present. They are known collectively as compass error.

The drawings below illustrate what can happen to a boat which does not allow for compass error and how such error can be applied so that the boat may steer her required course.

Let's assume that we are fishing on an offshore reef when the weather turns thick and we are unable to see the shoreline. We know that the course back to port is 270° (taken from the chart) and we have a compass error of, say 15°W. If we do not allow for the error, but steer the course 270° on the compass, the error will

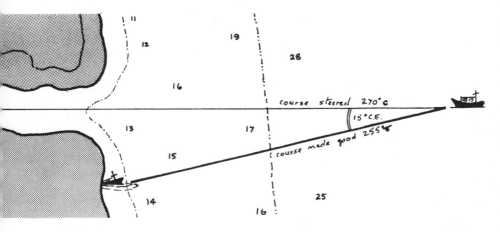

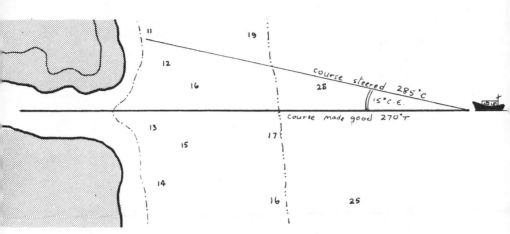

push the boat 15° off course and she will wind up wrecked on the coastline to the south of the harbor entrance.

By allowing the compass error the boat, although steering 15° to the north of the harbor entrance, will be pushed back 15° by the error and make a true course of 270° right into the harbor.

The total compass error affecting the boat is computed by adding or subtracting the variation to the deviation as follows:

LIKE NAMES ADD: UNLIKE NAMES SUBTRACT

Thus:

Variation 10°E Deviation 2°E Compass Error 12°E
Variation 10°E Deviation 2°W Compass Error 8°E

Applying the compass error

Since it is not affected by any magnetic forces, everything on the chart relates to true north. By contrast, the courses steered and bearings used are related to magnetic or compass north. It follows that between the workings on the chart and applying those workings in practice to the boat's compass, the compass error must be allowed.

Think of it this way: the chart is kept below decks on most yachts. The compass is up on deck. Thus every time the navigator goes up or down the companion way between the deck and the cabin, he must make allowance for compass error to any navigational workings. Making this a habit can eliminate the very easy possibility of forgetting the error and placing the boat in danger.

45

Unless compass error is allowed correctly, the boat will be steering many degress off course.

Applying the compass error can be done in a number of ways, but the easiest way to avoid confusion, and one which is absolutely foolproof, is to remember the jingle:

ERROR EAST, COMPASS LEAST—ERROR WEST COMPASS BEST

An example is probably the best way to illustrate the use of this jingle:

Variation 10°E   True course on the chart 269°T
Deviation 3°W
(error east, compass least)
Error 7°E   Course to steer by compass 262°C
Variation 10°E   True course from chart 269°T
Deviation 17°W
(Error west, compass best)
Error 7°W   Course to steer by compass 276°C

Be very careful when applying the error to compass bearings as opposed to courses. Because this time we are coming down the companion way from the compass to the chart and the system appears to be reversed:

46

Variation 10°E Bearing taken on compass 167°C
Deviation 5°E
(error east, compass least)
Error 15°E Bearing to lay off on chart 182°T

It is a wise precaution always to indicate chart bearings and courses with T (true) and compass bearings and courses with C (compass). This prevents any confusion which might otherwise arise.

Swinging for compass

It sometimes happens that the services of a compass adjuster are not available to find the deviation. When this is the case the navigator must do it himself by a procedure known as swinging for compass. The following are the steps that should be followed:

1 Locate two range objects (objects in line) and determine their true bearing on the chart. A single object is only acceptable if it is more than six miles away.
2 Apply the variation to the true bearing to find what is known as the magnetic bearing.
3 Secure the boat to an anchor or a mooring buoy so that the objects are exactly aligned and their bearing does not change. By means of the motor or a rowboat, swing the boat's head until it is pointing due north.
4 Read off the bearing of the range on the compass.
5 The difference between the compass bearing and the magnetic bearing is the deviation error when the boat is on a north heading.

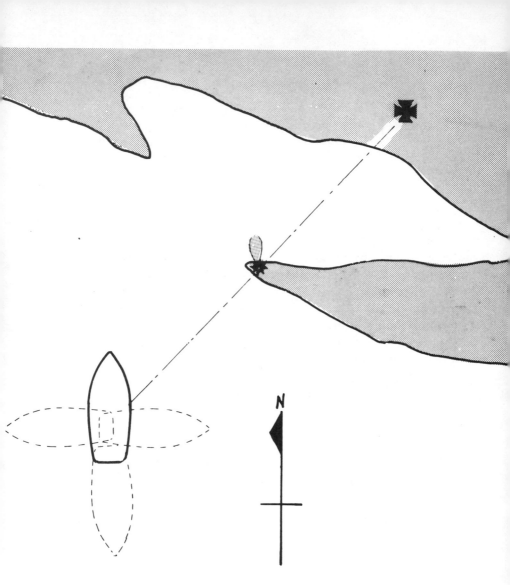

6 If the true bearing is greater, the deviation is named east, if the
   compass bearing is greater the deviation is named west.

7 Repeat the procedure with the boat heading on each of the
   cardinal points of the compass (N,S,E,W), and repeat again
   with the boat heading on each of the intercardinal points—
   NE,SE,SW,NW.

8 From the results, make up a deviation card as illustrated.

*Note:*Because the electrical circuits of the motor create their own deviation error, it is a wise precaution to carry out this procedure first with the motor running and then with it stopped.

Chapter 4 Lights and Lighthouses

Lights are the navigator's signposts. Without them the job of keeping track of the boat's progress along the coastline would be much more difficult. By day the lighthouse buildings offer readily identifiable objects and for this reason are usually painted white. By night, the light they contain offers almost the sole visible means of navigation on otherwise dark stretches of coastline.

Types of lights

There are three principal types of lights used for coastal navigation:

1 *Primary seacoast lights*, which are usually high-powered lights in fixed structures placed at strategic points along the coastline.
2 *Secondary lights*, are those which help guide vessels into harbors, estuaries and other waterways.
3 *Minor lights*, are of moderate to low intensity mounted on fixed structures which help to guide the navigator through the shoals and hazards of inshore waterways.

The primary seacoast light

This is a long range light, usually white in color and showing over a wide arc of the horizon. Its purpose is mainly to assist navigators making offshore passages up and down a coastline or to guide them towards the proximity of a harbor entrance. Its range varies, but 20-25 nautical miles is fairly standard. Each light is usually strategically placed so that as one is left behind, the navigator making a passage along the coastline is not left too long in the dark, as it were, before another appears over the horizon.

To assist identification, these lights have very different characteristics. Although a few may be colored, as a general rule color inhibits their range and for the most part they show a white light. This light is divided into flashes or groups of flashes with differing time cycles so that one light will not be confused with another (see 'characteristic of a light', later in this chapter).

For ease of identification by day the lighthouses are usually placed right on the edge of the coastline or on an offshore island, and are distinctive in shape and color. Stripes and squares are sometimes used to differentiate one from another where they are close together. But where they cannot be confused, the classical tall white 'needle' building is most common.

Secondary lights

Usually placed on headlands at the entrance to harbor or some other strategic position, these are lights that take up where the big primary seacoast lights leave off. They usually mark the approaches to a harbor, river or channel so that the offshore navigator, having made a landfall on the seacoast light, can pick up the secondary light and make his way towards the estuary or harbor entrance.

These lights are usually of fairly intense power, although less powerful than the seacoast light, and often use colors, either as a warning of inshore dangers or as a guide into the channels. Red, green and white are the most commonly used colors with red a particular favorite to indicate dangers.

The structure bearing the light may be anything from the classical lighthouse building to a small iron framework on top of which sits the light itself.

Minor lights

These are the lights which mark the channels, shoals and reefs of navigable harbors and estuaries. They are usually mounted on piles or metal structures although they can take any form or shape. Much will depend on the importance of the channel they mark and its use for commercial shipping.

Minor lights are principally green, white and red but other colors such as blue, may sometimes be used. They have a very limited range, often only a mile or so, and are invariably unattended. Most conform with the characteristics of the lateral system of buoyage (see Chapter 9).

Navigation by day is relatively easy, but if entering a port at night the navigator must be certain of his lights.

Making port

The offshore navigator, making into a harbor at night, will follow a routine somewhat similar to that followed by an aircraft pilot on approaching an airport. Just as the aircraft homes in on a major beacon near the city, so the boat navigator uses the primary seacoast light to guide him towards the entrance to a port. An aircraft closing in on airport is guided into the approach pattern by the control tower, and similarly the boat navigator uses the guiding beacons of secondary lights to make his approach to the harbor entrance. As the aircraft touches down between two rows of runway lights so the boat is steered to her berth through rows of minor harbor lights, supplemented by those of the buoyage system.

The loom of a light

Long range seacoast lights usually depend on the concentration of *Fresnel* lenses to condense their light into a strong beam. These lenses rotate mechanically, thus sweeping the beam around the horizon like a probing searchlight. As the beam passes a boat offshore, it is seen from the boat as a brilliant flash, again like being swept by a searchlight.

On clear nights, long before the actual light appears over the horizon, the sweeping beams can be seen in the sky ahead. This is known as the 'loom' of the light, and apart from allowing early identification of the light itself, can be used for plotting the boat's position (see Chapter 6).

The break of the light

At the moment that the light comes over the horizon the loom gives way to the light itself—just like a car headlight coming over a hill. Instead of the sweeping beams of the loom the light appears as a brilliant flash. This is known as the 'break' of the light.

Range of a light

Alongside the light symbol on a chart is the abbreviated characteristic which indicates its performanance. Included in this characteristic is the range of the light in nautical miles.

Obviously, the light will be seen at differing distances from different heights above the deck of the boat. For the range marked on many charts, a standard 'height of eye' of about 4.5 meters (15 feet) is used. For navigators with a height of eye above sea level other than 4.5 meters, the range of the light will vary accordingly.

A table called the *Geographic Range Table* enables the range of any light, seen at any height of eye, to be calculated. (Bowditch, Table 40)

The range on United States charts published after June, 1973, is the actual distance that the particular wattage of the light will penetrate clear atmosphere; the *nominal range*.

Height of a light

The height of a light is also included in the details carried on the chart. It has a number of uses in navigation. To allow for a safety margin, heights on a chart are all measured above *mean high water springs*, (MHWS).

Characteristics of a light

The characteristic of a light is the make up of its flashes, color and

The loom of a light.   **The break of a light.**

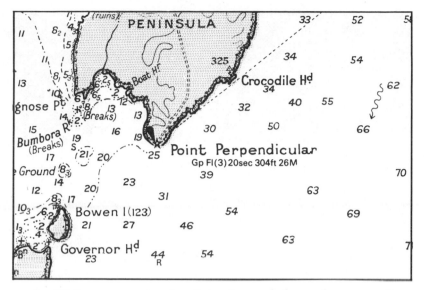

cycle. With many navigation lights located along a busy coastline, (to say nothing, of course, of shore lights) confusion would result if many of the lights had similar characteristics. In order to prevent such confusion no two navigation lights with similar characteristics are located together on a stretch of coastline.

There are three principal flash characteristics used for navigation lights. They are given below together with the abbreviation used to describe them on the chart:

F   Fixed   A constant light with no break.

Fl   Flashing   The period of light is shorter than the period of darkness.

Oc   Occulting   The period of light is longer than the period of darkness.

These basic characteristics can be extended by forming the

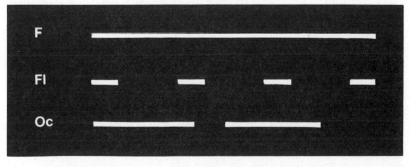

flashes or the occults into groups. The light is then said to be Group Flashing or Group Occulting and the number in each group may vary by up to four. Thus:

GpFl(3) or Fl(3)   A group of three short flashes followed by a period of darkness.

Gp Oc(2) or Oc(2)   A group of two long occults followed by only a   brief period of darkness.

To prevent confusion even further, colors may be added to the charactertistic of the light in which case the color will be indicated by the inclusion of its capital letter. Where the light is solely white no capital letter is used.

Fl (4) R   A red light flashing in groups of four.

To make even more certain there is no confusion, especially when white is the only color used, the characteristic of the light is given a time cycle totally different to any other light in the area. The time is given in seconds and includes the *whole* cycle of the light.

Where group flashing or occulting is concerned, the timing of the cycle commences at the beginning of the first flash of one group and ends at the beginning of the first flash of the next group. Thus:

Fl(3) 10s   A group flashing light with three short flashes followed by a long period of darkness, the whole cycle taking ten seconds.

There are far too many different light characteristics to cover them all in a book of this type. A publication produced by the U.S. Coast Guard and called simply *Light List* gives full details of all coastal, intracoastal and inland waterway lights that affect boat navigation in U.S. waters. It goes without saying that such a publication is vital to safe navigation.

In order to make them easy to identify, lighthouses are painted white or with distinctive colour patterns.

Sectored lights

A popular form of coastal light used for leading boats into an estuary or harbor, or for marking offshore hazards is the sectored light. This uses two or more colors, divided into sectors to indicate to the navigator just where he is in relation to dangers.

As a general rule, the white sector is the safe one, but this should be checked on the relevant chart where the sectors are indicated by pecked lines. A navigator entering a harbor in the white zone and seeing the light turn red or green may be sure he has wandered from the safe channel, and should immediately check his position. Since the different sectors are obtained by placing colored screens around the light, the flashing or occulting characteristic of the light is the same in each zone regardless of its color.

Some seacoast and secondary lights use colored sectors to indicate offshore dangers, and once again reference to the chart will indicate the nature of the problem and the color of the light which covers it. It is not uncommon for the main light to be separate from the colored sector which might otherwise inhibit its range. In this case both characteristics are shown on the chart.

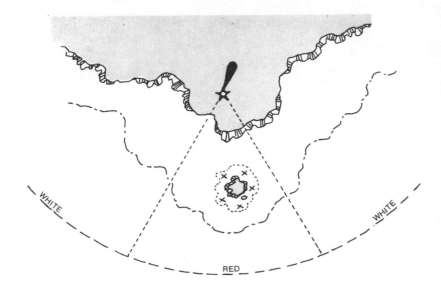

Range lights

A very accurate method of leading a boat in through a narrow or
difficult entrance is the system known as range lights. This
is used widely in canals or rivers where the broader beams of
sectored lights are not sufficiently accurate for the fine navigation
required.

Range lights comprise two individual lights, usually colored,
and placed so that the lower light is some distance in front of the
upper light. These lights are located in such a position that on
entering the channel the navigator sees both lights, one above the

other. Any tendency for the boat to stray to the right of the channel will cause the lower light to move to the left of the upper and vice versa. To keep his boat in the centre of the channel the navigator must keep both lights exactly in line, one above the other.

By day brightly colored markers replace the lights and the same procedure is followed. This is a particularly good system where channels are subject to change due to shifting sand bars or river floods. By moving the lower light, the two lights can be realigned to the new direction of the channel. It is a system widely used to mark the entrance to a harbor over a dangerous bar.

Chapter 5   **Planning the Coastal Passage**

A passage along a coastline must be carefully planned beforehand so that the boat makes the best possible track consistent with safety. The best track may not always be the shortest, since it may be advantageous to divert in order to gain assistance of a favorable current or to avoid a dangerous area in unfavorable weather conditions.

But as a general rule, the best track is the shortest consistent, as mentioned, with safety. So assuming that there are no hazards *en route*, the planned course from the point of departure to the port of arrival will be a straight line. If there are dangers or hazards on the way, then the best track will probably be a composite course made up of shorter courses designed to pass a safe distance off the dangers.

Departure! And the excitement of a coastal passage to look forward to.

Arrival off the destination port after a successful passage.

Arrival and departure positions

Trying to lay off a series of courses for a boat to follow through the intricate channels of a harbor or estuary is virtually impossible. Avoiding collision with other vessels is just one way such planned tracks would be upset.

So generally speaking, any navigation within the confines of a harbor is visual, using channel markers and buoys to make the best path downstream and out into the open sea.

Once clear of the restrictions of sheltered water, however, a track can be planned along the coastline with little risk of interference. For this reason, all coastal navigation planning commences at a point *outside* the port of departure and terminates at a point *outside* the arrival port. Such points are called, respectively, *departure* and *arrival* positions.

It is from the departure position that navigation along the coast begins. The distance log is set at zero and all calculations are commenced at this point. A fix is made of the boat's position when on her departure point (see Chapter 6), and the passage recorded as starting at that time.

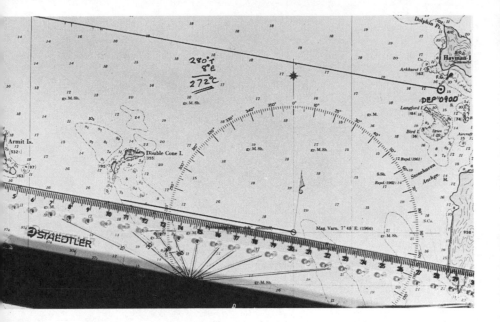

Laying off a course

With the departure and arrival positions established and plotted
on the chart, and presuming that one course can be steered for the
full length of the passage, finding the course to steer is a simple
process:

1 Join the departure and arrival positions with a straight line.
2 Place the parallel rules along this line. Carefully transfer it
   through the center of the nearest compass rose.
3 At the point where the edge of the rules cuts the circumference
   of the compass rose *in the direction in which the boat is to travel*, read
   off the true course to steer.
4 Convert to a compass course by applying the compass error.
   This is the course to steer by compass between departure and
   arrival points, and to make good the track laid off on the chart.

Safety circles

If the course laid between departure and arrival positions is clear
of all hazards, then no problems arise. But if there are dangers *en
route*, or if the course has to round a headland or island, then the
navigator must determine beforehand how far off such dangers he
wishes to pass.

   Much will depend, of course, on conditions. Rounding a rocky island in daylight and in clear weather, for example, it may be safe to pass within half a mile. The same hazard in foul weather or at night may require a clearance of two or three miles. Only the navigator's skill and experience can determine what is a safe distance under the prevailing conditions.

Having decided on this safe distance, a 'safety circle' is drawn around the danger as follows:

1 Set a pair of compasses with radius the safe distance determined.
2 With the point of the compasses on the outermost part of the danger, draw an arc of a circle to seaward of it.
3 Providing the boat stays outside this circle she will always be more than a safe distance off the hazard.
4 From the departure position, draw the courseline at a tangent to this safety circle. This is the course to clear the danger by the predetermined safe distance.

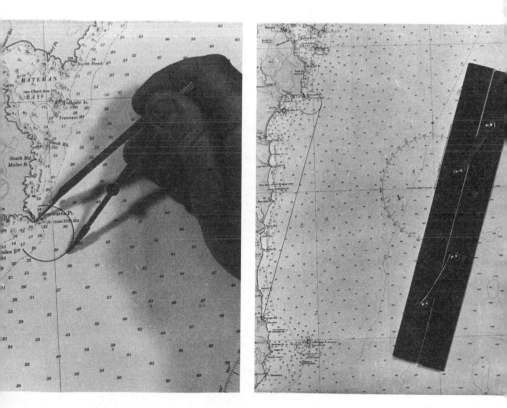

Laying off a composite course

Where the course between departure and arrival positions cannot be drawn as one straight line due to intervening hazards, the following procedure is adopted:

1 Draw a safety circle around the first hazard.
2 Repeat with each hazard the boat has to clear before reaching her destination.
3 From the departure position draw the first course at a tangent to the first safety circle as described.
4 Join the first and second safety circles with a line at a tangent to each.
5 Repeat the procedure joining all the safety circles with tangents until the last circle is reached.
6 Draw the final course to the arrival position from a tangent to the last safety circle.
7 Read off the various true courses on the compass rose, and convert to compass courses by applying compass error. Write each course on the chart against each courseline.
8 These are the courses to steer in order to make the shortest passage between departure and arrival positions while passing all dangers at the predetermined safe distances.

Fog, rain and snow (bad visibility)

It goes without saying that in planning his boat's passage, particularly with regard to passing a safe distance off hazards and dangers, the navigator assumes the weather will be fine and clear. However, in the course of sailing that passage, the weather may deteriorate and allowances for a safe distance off hazards will need to be increased.

Since no navigator, no matter how experienced, can predict exactly what the weather is going to do, it is more than likely the

A composite course.

Navigation is relatively easy when conditions are good.

Sailing close to a lee shore is always dangerous

entire passage will need to be redrawn if bad weather is encountered. Just how much extra clearance will be needed will depend
on just how bad the weather is, how well the navigator knows the
coast, and what sort of coastline it is. Experience is the best guide.

Lee shore

Sailboats, in particular need to give a wide berth to a lee shore.
This is the shore onto which the wind blows, and navigating too
close can mean that in the event of an accident (lost mast, motor
failure), the boat will blow towards the shore.

Since it is hard to predict which will be the lee shore when
planning a passage some days beforehand, this is another adjustment which may need to be made when under way along the
courselines. The safety circle which covers a hazard to leeward
should have a greater margin of safety than one which covers a
danger to windward.

Night navigation

There is always less opportunity to keep a check on the boat's
progress at night since many of the shore objects, and often the
shore itself, cannot be seen or can only be seen indistinctly. Safety
allowance for hazards and dangers should therefore always be
much greater at night than during the daytime.

Instruments are useful when navigating at night

Heavy weather

As with night navigation, visibility is often restricted in bad
weather and precautions should be taken to ensure that safety
margins are increased. Heavy weather can cause the boat to be
buffeted off course very easily and make recovery very difficult.

Leeway

The effect of the wind pressure on a boat under way causes her to
drift off her courseline. It is known as leeway. On a motor yacht it
can be fairly negligible, but on a sailboat it is considerable.
Indeed, leeway of up to 10° is not at all uncommon, and if the boat
is to hold an accurate course, this must be taken into consideration
when planning the passage.

There is only one way to find how leeway affects your boat, and
that is by experience. It can not be calculated, other than perhaps
by computer, and few navigators have one of those aboard. By
sailing directly for a shore mark and noting the amount of drift to
leeward, an idea of how much leeway she makes under differing
conditions can be assessed.

The most difficult aspect of calculating leeway is the number of
factors involved. Leeway is different, for example, when sailing
into the wind to when sailing before the wind. Similarly it is
different when the boat is sailing upright to when she is heeled on
her lee gunwale.

When a fair idea of leeway has been gained, it can be allowed
when setting a course by applying the amount of leeway (the
leeway angle) *into* the wind.

Tides and currents

Currents and tidal streams create some of the biggest problems encountered on a coastal passage. An ocean current runs consistently in the same direction, a tidal current flows backwards and forwards, usually twice a day.

Ocean currents are mostly associated with ocean navigation, but in some parts of the world, where they move close to the coast, they affect coastal navigation. Tidal currents are a feature of coastal waters and will be encountered in every estuary and harbor as well as some distance offshore.

Tidal current tables and charts are published by The National Ocean Survey for waters where tidal streams are strong and vary considerably in direction. These usually indicate the flow of tidal streams at hourly intervals before and after high water. Information concerning both tidal and current streams can be obtained from the *Sailing Directions* or *Coast Pilots*. When the direction and rate of flow of a current is known it can be counteracted as described in Chapter 7. But many tidal flows are irregular and many currents have eddies which swirl in all directions.

This means that the effects of tidal streams and currents on the boat often cannot always be predicted, adding yet another problem to the many encountered by the navigator when making a coastal passage.

Latitude and longitude positions

Although the boat's position is not often plotted in co-ordinates of latitude and longitude when making a coastal passage, it is some-

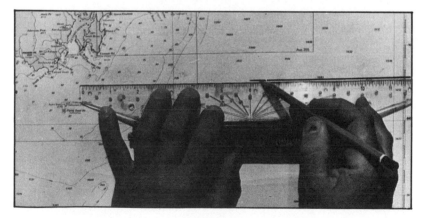

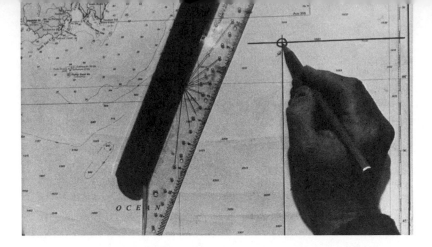

times necessary to locate a position given in latitude and longitude on the chart. The procedure then is as follows:

1 Locate the latitude position on the left or right side of the chart and mark it with a pencil.
2 Place the parallel rules against the nearest parallel of latitude on the chart then run them down until one edge is touching the latitude marked. Draw a line across the chart.
3 Repeat the procedure using the longitude scale and the nearest meridian of longitude.
4 Where these two lines intersect is the plot of the latitude and longitude position.

It follows that by reversing this process the navigator can find the latitude and longitude of his own boat's position. Latitude and longitude positions are usually entered in the log book as a record of the boat's progress during the passage.

DR and EP

The DR (dead reckoning) position of the boat is not a positively established position. It is a position arrived at by calculating the boat's progress along her past courses and log distances. The usual method of finding the DR position when coasting is to mark off along the courseline steered, the distance run by log from the last known fix position. A DR position is marked with a triangle or half-circle. The dot in the center marks the exact spot.

An EP (estimated position) is found in a similar manner to a DR position, but with the application of any known tidal current effect. An EP is therefore a more accurate position than a DR position. It is marked on the chart with a triangle in the same way.

Chapter 6   **Plotting the Track**

Set and drift

Plotting is the means whereby the navigator keeps track of the boat's progress. Once the course is laid on the chart and the boat begins her passage along it, a number of unknown factors will affect her and probably push her off the planned courseline.

These factors, some of which were described in the previous chapter, can include tidal effects, currents, wind and wave effect—in short, anything which the navigator does not know about and therefore has not allowed for. Even bad steering on the part of the helmsman can add to the factors setting the boat off course.

The sum total of all these unknown factors is called *set*, and the distance it pushes the boat off course is termed *drift*. These terms are also used in connection with unknown currents and tidal flows in exactly the same context. The set of a current is the direction it pushes the boat. The drift is the distance it pushes her off course.

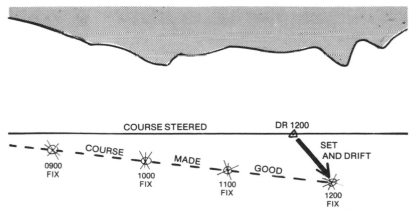

69

The timing of plots can be varied according to conditions, providing they give the navigator a clear picture of the track the boat is making and the forces affecting her.

Timing of plots

It follows, then, that if an unknown set is affecting the boat she will not follow the planned courseline, but will be pushed either inshore or offshore. In practice a boat rarely keeps to her track but is affected by set on almost every course. Later we shall deal with the question of determining the amount of set and drift and allowing for it in our calculations. But initially the important thing is to keep track of what it is doing to the boat and ensure that she is not pushed into any danger.

This is done by plotting her position at regular intervals. Just how regular depends on her location and the conditions at the time. Close inshore on a dangerous coastline, for example, it would be wise to make a plot every half hour. Well offshore and in clear weather, once an hour would suffice.

The fix

A fix is an exact plot of the boat's position determined by navigational means. Plotting the boat's track is done by means of a series of fixes, indicating her progress along the courseline. Most coastal fixes are the result of taking compass bearings of shore objects. There are a number of different methods of obtaining a fix, each of which depends on the number and type of shore objects visible.

Laying off a bearing

In Chapter 3 we took a bearing of a shore object using the hand-bearing compass. By laying off (plotting) this bearing on the chart

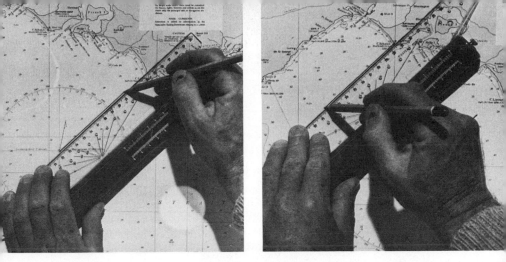

we are well on the way to determining the boat's position, for if a bearing is correctly taken and laid off on the chart, then:

The position of the boat must lie somewhere along that bearing.

As described, the bearing is taken by sighting over the compass. It is then converted to a true bearing so that it can be laid off on the chart. The procedure for laying off is as follows:

1 Find the true bearing on the circumference of the compass rose nearest the object.
2 Place the parallel rules so that one edge runs through the centre of the compass rose and cuts the circumference at the bearing.
3 Carefully move the parallel rules across the chart until one edge touches the object from which the bearing was taken.
4 Draw a line along that edge of the rules to seaward of the object. This represents the bearing taken of the object, and the position of the boat must lie somewhere along this line.

Because a fix of the boat's position may be made by use of these compass bearings, it follows that they must be taken with great care and accuracy. An error of one or two degrees in the reading can be magnified into even larger errors when plotted on the chart and provide a completely false position.

Typical objects for taking bearings

As mentioned, literally anything on shore that is readily visible from sea (and is marked on the chart) can be used for taking bearings. However, by virtue of their shape or location, some objects are better than others. A lighthouse, for example, since it is located right on the coastline, is painted white and easily pinpointed, makes the ideal bearing object. A mountain, which may be situated way back from the coast, has no definite peak and may be confused by inshore haze, is not good.

The cross bearing fix

This is the most commonly used form of fix, particularly in daytime. It requires at least three shore objects to be readily visible at the same time. Ideally, these shore objects should be spaced at angles of roughly 60°. The finer (or broader) the angle between each object, the less accurate the fix.

The type of object is not important providing it gives a good bearing and is easily identified on the chart. Two bearings can be used, but the margin for error is greater and three is accepted as offering the best fix. The procedure is as follows:

1 Select three shore objects well spaced and identify them on the chart. Take a bearing of each as quickly as possible.
2 Convert each bearing to true and measure the first against the compass rose.
3 Transfer the bearing through the shore object and lay it off on the chart.
4 Repeat the process with the second bearing.
5 And the third bearing.
6 Where all three bearings cross is the fix of the boat's position.

Cocked hat

It frequently happens in small boat navigation that the three bearings of a cross bearing fix do not cross exactly, but form a small triangle. This is known as a 'cocked hat'. Because of the difficulties in taking three quick bearings from a tossing boat, the cocked hat can be accepted as a fix providing the triangle so formed is not too large. A big cocked hat would be an indication

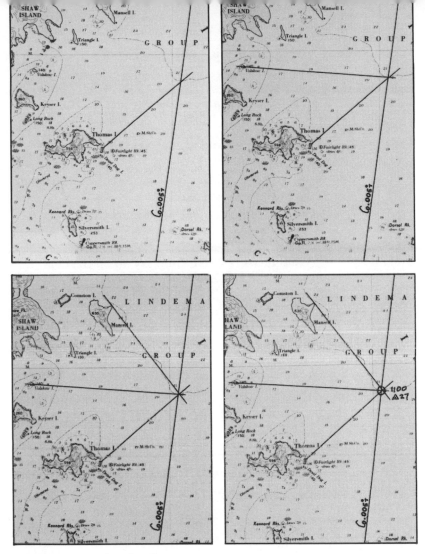

Plotting a cross bearing fix.

of errors somewhere in the fix and bearings should be taken again.

If for some reason it is not possible to retake the fix, the corner of the triangle closest to shore should be taken as the fix position. This provides a margin of safety.

The running fix

It is not always possible to obtain three shore objects close enough to use for a cross bearing fix. Indeed, often there is only one suitable object visible, particularly at night when fixes can be made

73

only with lights, and navigation lights are usually well spaced out.

In this case, the fix used is called a running fix, also sometimes called the transferred bearing fix. It again involves the use of compass bearings, but also requires the reading of the distance log. The procedure is as follows:

1 Take a bearing of the object and at the same time take a log reading.
2 Convert the bearing to true and lay it off on the chart. Mark the time and the log reading alongside.
3 Allow the boat to proceed on course for some distance, say three or four miles, then take another bearing of the same object and another log reading.
4 Convert this second bearing to true and lay off on the chart.
5 From the point where the first bearing and the courseline inter- sect, mark off along the courseline the distance run by log bet- ween the two bearings. Call the terminal point X.
6 Place the parallel rules against the first bearing and transfer it carefully to point X.
7 Draw a line through X.
8 The point where the first bearing, transferred through X, cuts the second bearing is the fix of the boat's position.

Because this fix involves a run between bearings, there is a small possibility of error creeping in. Thus, one running fix is considered only a guide to the boat's position, and the process is repeated several times as the boat passes the object. After two or three running fixes, all errors will be eliminated and the boat's position will be established beyond doubt.

Running fix between two objects

Where two well-spaced objects are not suitable for a cross bearing fix, the running fix may be adopted. The bearing of the first object is transferred along the courseline for the log distance run in just the same way as when using one object. Where this transferred bearing cuts the bearing of the second object is the fix of the boat's position.

The cross bearing fix with three shore objects, and the running fix when only one shore object is visible, are the basis on which plotting a boat's progress along her coastal passage is established. However, there are other useful fixes which can be used providing conditions are suitable and objects are available.

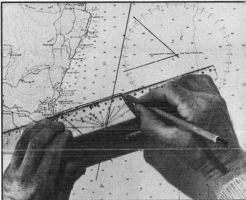

Stages in plotting a running fix.

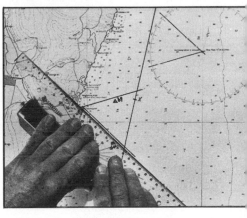

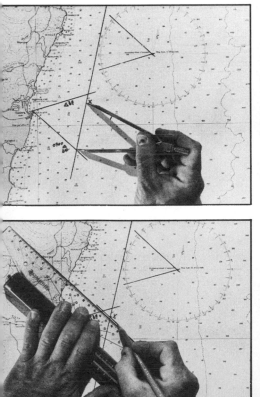

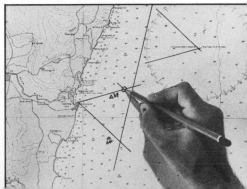

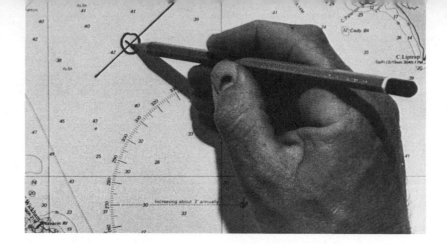

The extreme range fix

This is an excellent fix, used mostly at night when making a landfall, or approaching a lighthouse from a long distance off. It depends on clear weather and seeing the 'loom' of the light (see Chapter 4). The procedure is as follows:

1 Watch the loom of the light. Identify the light on the chart.
2 When it 'breaks' take a bearing. Convert to true and lay off on the chart.
3 Enter the Bowditch *Geographic Range* table (Table 40) with arguments height of eye of the navigator, and height of the light (from the chart). Take out the 'distance off' the light.
4 Measure this distance along the bearing to seaward of the light.
5 The result is a fix of the boat's position.

Four point bearing fix

This is a handy rule-of-thumb fix which does not require the use of compass or chart. It is ideal for checking the boat's position in clear weather and when all things are bearing an equal strain. In short, it is a simple but effective check of the boat's progress and whether or not she is being pushed inside or outside the courseline by an unknown set.

The four point bearing is the bearing which is 45° to the course on either side of the bow. Generally speaking, the points of the compass are not used greatly nowadays. But with dead ahead as zero, it was always accepted that the beam of the boat (90° from ahead) was eight points on either side and the intervening area was divided up into points of 11 1/4° each.

Although more accurate using a hand-bearing compass, the four point bearing fix can be obtained without any instruments other than the log.

However, only the four point position, 45° on either side of the bow, is used in modern navigation, the remainder of the points having gone the way of the cardinal compass which gave rise to them.

A good navigator will know the four point bearing on his boat. From the cockpit position, a stay, ventilator, or even a mark on the gunwale will indicate the 45°, four point location from the boat's head. With this established beforehand, the procedure is as follows:

1 Watch the approach of the object until it is on the four point bearing. Read the log.

2 Allow the boat to continue along her courseline until the object is on the beam bearing. Read the log again.

3 The distance travelled by log between the two bearings is equal to the distance off the object on the beam bearing. The navigator, knowing how far off the object he intended to pass, will thus know whether or not the boat is maintaining her courseline.

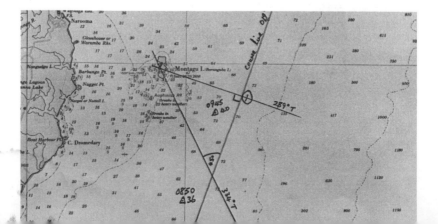

Doubling the angle on the bow

This is another handy check on the boat's progress, requiring the use of both compass and log. The procedure is as follows:

1 When the object is fine on the bow, take a compass bearing. Read the log.
2 Find the *relative bearing* (i.e. the angle on the bow) by adding or substracting the bearing to the course being steered.
3 Allow the boat to continue on course until the relative bearing has doubled. Read the log.
4 The distance run by log between the two bearings is the distance off the object on the second bearing.
5 Convert the second bearing to true and lay off on the chart. Mark off along this bearing the distance run by log.
6 The result is the fix of the boat's position.

Fix by vertical sextant angle

This is perhaps the most accurate fix to use when 'rock hopping', or running close in to the coastline. It requires the use of a sextant as well as the hand-bearing compass, but is more accurate than any other fix and therefore ideal for plotting under close or difficult navigational conditions. Most experienced yachtsmen who use this fix prefer it even to cross bearing or running fixes when navigating close to the shore.

1 Take a bearing of the object, which needs to be reasonably high.
2 Take a sextant reading, bringing the top of the object down to sea level in the telescope.
3 Convert the bearing to true and lay off on the chart.

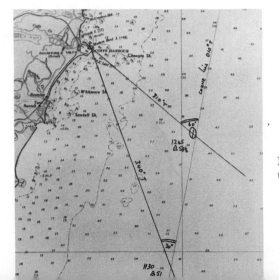

Doubling the angle on the bow.

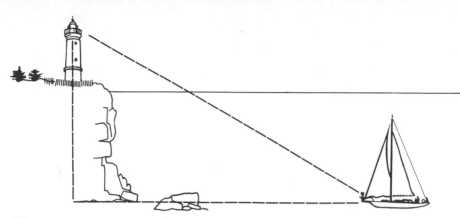

The sextant measures the angle of the lighthouse above sea level. The table converts this angle to a distance off the shore.

4 Enter the nautical tables (Bowditch, Table 34) *Distance by Vertical Angle*, using the arguments of the height of the object from the chart and the angle measured on the sextant. The result will be the distance off the object.

5 Measure with the dividers this distance and mark it off along the bearing laid off on the chart.

6 The result is a fix of the boat's position.

Ad infinitum

There are, of course, many other ways of fixing the boat's position and keeping a check on her progress. Indeed, it depends purely on the navigator's ingenuity as to the number and type of fixes which can be used for coastal navigation.

Chapter 7 Countering Winds and Currents

Set

As mentioned earlier, set is the factor usually created by tidal current or flow, which pushes the boat off course. It is caused by a number of factors, some known and some unknown. It can be allowed for in cases where its effect is known, but where it is composed of unknown factors its effect must first be found before action to counter it can be taken.

Countering a known set

When laying a course across a stretch of water which contains a known movement such as an ocean current or an established tidal current, the navigator can take action to counter the effect of the set which will be experienced, and keep his boat on the planned courseline. The procedure is as follows:

1 Lay off the planned courseline from departure to arrival positions.
2 From the departure point, lay off a line representing in both direction and distance one hour of the current. Call the terminal point X.
3 Set the dividers at a distance representing the number of miles the boat will travel in one hour without the effect of the current.

Close inshore, currents are usually tidal and predictable.

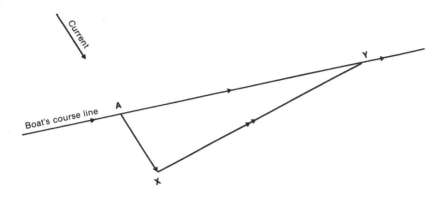

4 Place the point of the dividers on X and swing them until they intersect with the original courseline at point Y.

5 Join X and Y. This line, when measured on the compass rose, is the true course to steer to counteract the known current and maintain the original courseline.

6 The distance along the courseline from departure point to Y represents the distance the boat will travel in one hour while being affected by the current.

The one hour period is arbitrary, of course, and where the scale of the chart is small, two or more hours can by used, providing the same period is used for both current and boat.

Countering an unknown set

This is much more difficult in practice because before countering an unknown set, it's effect must first be found. However, this is a common practice when coasting as most of the set encountered will be due to tidal currents which are often erratic. Only on a few occasions will the effect of set be known accurately enough to use the previous method.

The course made good

As the boat progresses along her courseline, the set will tend to push her inshore or offshore from the courseline. At first this will be a relatively small amount, but will be exaggerated as time passes, to the point where she has drifted so far off course that adjustments must be made to get her back on course for her original destination.

Since plotting is carried out at frequent intervals, a series of fixes on the chart will show the amount and direction the boat has

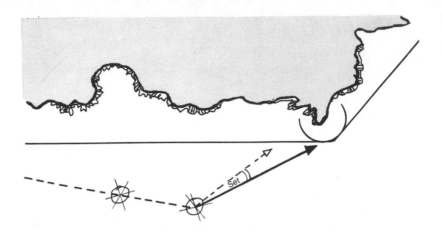

When all other factors have been allowed for, the newly-found set must be applied to the compass course to steer to get the boat back on her original course.

drifted off course. If these plots are joined by a line from the departure position (or the last known fix) the track she has made as a result of the influence of an unknown set can be seen. This is known as the Course Made Good (CMG) and can be measured on the compass rose. The distance sailed along the CMG is known as the Distance Made Good (DMG).

The angle of set

If the boat has been steering one course and making good another, then the difference between them is the effect of the unknown set. It is measured as an angle simply by subtracting the course steered from the course made good (or vice versa) and is termed the angle of set.

Correcting the course

By the time the angle of set has been established, the boat will have been pushed well off her courseline. Thus a new course must be laid off from the last fix to her destination or arrival position. This is measured on the compass rose and must be adjusted for compass error before altering the course to steer. However, while this new course will head the boat back to her original destination there is still the effect of the set to take into consideration. This is done by applying the angle of set to the new course against the direction of the current.

Unfortunately, this newly found set is likely to change as the boat proceeds along her courseline, for rarely is a set constant for long periods. By continuing to plot the boat's progress with a series of fixes at regular intervals, a picture of what is happening to her as a result of the changing set becomes apparent, and adjustments can be made when required to keep her travelling along her original courseline.

In practice it is often unnecessary to re-lay the course and find the angle of set each time the boat drifts from her courseline. By occasionally making small adjustments to the course steered, the smart navigator can keep the boat pretty well on her planned course the whole time, and avoid too many dramatic course changes.

Leeway

As described in Chapter 5, leeway is another of the great unknowns in coastal navigation. However, it can be determined with reasonable accuracy during the course of sailing the boat over a long period of time. Then the angle of the leeway can be measured and applied to the course steered in much the same way as the angle of set.

The only real way to find leeway is to sail the boat under all conditions and rigs, remembering that even one sail change can affect the leeway, as can the amount of heel and the angle to the wind at which she is sailing. The procedure is roughly as follows:

1 Establish a buoy or similar mark in the water as the starting point. Determine the compass course to steer from this point to another buoy or shore object, by laying it off on the chart then applying compass error.

2 From the starting point, sail the boat along this course, *steering by compass.*

3 As the boat progresses, take a quick series of fixes to establish her course made good on the chart.

4 The difference between the course steered and the CMG will be the leeway under the existing conditions.

5 Repeat with different weather and sail conditions until satisfied that you have established the boat's leeway on most points of sailing and under most moderate conditions.

Leeway should be found and checked as described before leaving port.

Note: It goes without saying that this method will only be accurate if carried out when there is no tidal current movement. The safest system is to make a number of runs along the course at slack water (high or low tide) in an area where tidal movement is accurately predictable.

Building up an accurate picture of the boat's leeway is a job that will take some time in order to encompass all the different points of sailing and different wind conditions. By using a little ingenuity, however, the navigator will find many opportunities during normal sailing, and particularly during racing, to use this method to check the leeway affecting the boat.

Applying leeway

As with set, leeway is easy to apply. The leeway angle is simply added or subtracted from the steered course by applying it *into* the wind.

Tack navigation

So far we have laid off courses, plotted and allowed for set and leeway assuming always that the boat can progress directly along her courseline. With a motor yacht this is always the case, but sailcraft cannot sail directly into the wind and are thus handicapped when making a course in that direction.

A fairly average tacking angle into the wind with a cruising yacht is around 45°. This means that there is a 90° zone, (45° on either side of the wind) in which she cannot sail. If the course to be steered is in that zone, then the boat must tack at 45° to the wind on either side of the courseline, using it as a 'mean' course by which to judge the length of the tacks.

If every boat sailed at exactly 45° to the wind, the navigational aspect would be fairly simple, but some boats 'point' higher into the wind than others, and then again, some point higher on one tack than another. What this means, in effect, is that the actual course that the boat will steer on her tack cannot be plotted ahead if she is to use best advantage of the wind, and sail on her best windward course. The procedure for tack navigation, then, is as follows:

1 Set the boat on her first tack, watching the sails to get her into the best tacking position. When she has settled down on this tack note the course on the compass.

2 Convert this course to true by applying compass error, and *take off the leeway* by applying it in the opposite direction to that used when putting it on (i.e. apply it in the direction the wind is blowing).

3 Lay the resultant course off on the chart. This is the tack she will make good providing the wind does not change direction.

4 At a predetermined point, put the boat onto the other tack and repeat the procedure.

5 Allow her to run back over the laid courseline, then at a predetermined distance on the other side, put her about again. Repeat the procedure in such a way that the boat's mean course into the wind is the course laid off on the chart.

If set has been allowed, it must be removed in the same way as leeway before plotting the tack course on the chart. Fixes should be taken continuously to check on the progress of the boat and adjustments made as they become necessary.

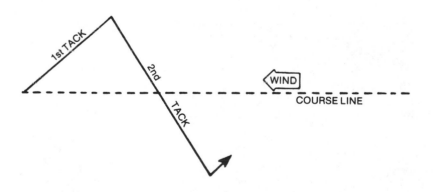

Chapter 8   Tides and Tidal Currents

Tidal current tables and charts

Many coastlines of the world have a simple ebb and flow of tide
with a moderate rise and fall, and relatively insignificant tidal
streams. However, many estuaries and land-locked coastal waters
are affected by strong tidal currents which can make navigation
very difficult. Not only does the speed of the flow vary at different
stages of the tide, but the direction of the flow can swirl in eddies
and create enormous problems for navigators unless they are
familiar with the area.

Tidal current tables and charts indicate the direction and
speed of tidal currents usually at hourly intervals before and after
high water. To find the direction and speed of tidal flow that will
affect his boat, the navigator should plot his anticipated position
across the chart at hourly intervals. In this way he can determine
the tidal problems he may expect while making passage. The tidal
current tables provide the time of slack water at both high water
(HW) and low water (LW) and the maximum current velocity in
knots at given reference points.

**Tides play a great part in harbor navigation—even when making an
anchorage.**

Ocean currents

Although primarily concerned with ocean navigation, when ocean currents skirt a coastline they have considerable effect on coastal navigation. Typical examples are the Aghulhas current along the east coast of Africa, the Gulf Stream where it runs along the coast of Florida and the infamous California Current on the West coast. Fortunately these currents are well charted in terms of direction and speed and, since they flow consistently in the one direction, their effects are easily predicted.

Tides

Tides are caused by a massive tidal wave that moves across the world constantly. Since some 75 per cent of the earth's surface is covered by salt water, this huge movement of the oceans affects literally every corner of every coastline in the world. As the tidal wave reaches a coastline it creates a rise in the water levels. As it moves on, the water level drops, thus creating the tidal cycle as we know it. The rising tide is known as the *flood tide* and the falling tide as the *ebb tide*. The average period of rise or fall for a tide is around 6 hours, but this can vary according to the geographical situation of the waterway.

Effect of sun and moon

The cause of tides is principally the gravitational effect of the sun and moon, the latter, being closer to earth, creating the stronger

Spring tides

Neap tides

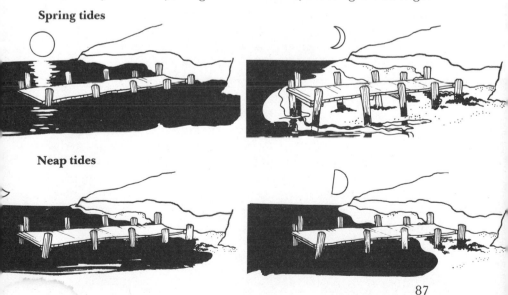

87

Tidal ranges are felt most in harbors and estuaries.

pull. This gravitational pull tends to draw the water around the earth's surface into a wave formation, and the rotation of the globe, in effect, moves this wave around its surface.

It follows that if the sun and moon are pulling in the same direction they will pile up a very high tidal wave and when pulling across each other, a relatively low wave. Because this wave builds up on either side of the world, the greatest pull of sun and moon is when they are on the same side of earth or on diametrically opposite sides. The time when their respective pull is negated by one another is when they are pulling at right angles.

Twice a month, at new moon (sun on the same side) and full moon (sun on the oppposite side of earth), this high tidal wave is built up. It is termed a *spring tide*. When the sun and moon are at right angles in the first and third quarter, the lower tidal waves create what are known as *neap tides*.

Spring tides

For the navigator, the effect of the tides comes in two forms—the flow of water created by the tidal current, and the depth of water over the sea bed created by rising and falling tide. Since spring tides are caused by a high tidal wave everything is exaggerated, so that the tidal flow is faster and the depths are greater. While the period of the tide is the same, its range is much greater, i.e. *high high tides and low low tides*.

Neap tides

These are the moderate tides. With a much lower tidal wave, the inrush of water is less and thus the tidal flow slower. Neap high

The depth of water over a dangerous bar is of utmost importance to the navigator. Chart datum indicates the least depth.

tides are generally less than average in height, and low tides are never very low. The period of the tide is again the same, but the range is much less, i.e. *low high tides and high low tides*.

Chart datum

The soundings marked on a chart are generally taken at the lowest tide of the month, and therefore there is *almost never* less water over the sea bottom than that marked on the chart. This level is termed chart datum, and tide heights are recorded above it. To find the depth of water over the sea bed at a given time:

the height of the tide from the tide tables is added to the depth of water on the chart.

Tidal anomalies

One of the most unsettling factors about tides is their habit of doing the unexpected thing at an unexpected time. Due to the topography of the sea bottom, it is not unusual for some ports to have an ebbing tide when the tide should be flooding and vice versa. In other ports a rise and fall (range) of tide of 40 feet may be experienced, creating all kinds of problems along the water-front. And to make life even more difficult, some ports have four high and four low tides a day!

Tide tables

Tide tables for different parts of the world are published by the National Ocean Survey and these are the only ones that should be used for navigation. Other tide tables, particularly commercial ones, may be quite accurate, but some are not, and using these can be risky when navigating in shallow waters.

The tide tables are drawn up for principal centers (known as *reference stations*) and secondary or *subordinate stations*. In the former case, the times and heights can simply be read off like any other table.

Subordinate stations, by contrast, have an adjustment of time and height to be applied to the readings of high and low water at a specified reference station. These adjustments are known as *tidal differences and constants*.

NEW YORK (THE BATTERY), N.Y., 1975

TIMES AND HEIGHTS OF HIGH AND LOW WATERS

JANUARY

DAY	TIME H.M.	HT. FT.	DAY	TIME H.M.	HT. FT.
1 W	0422	-0.8	16 TH	0431	0.0
	1043	5.1		1045	4.1
	1659	-1.1		1656	-0.2
	2321	4.6		2319	3.8
2 TH	0516	-0.6	17 F	0501	0.2
	1138	4.9		1120	3.9
	1749	-0.9		1722	0.0
				2357	3.7
3 F	0017	4.4	18 SA	0533	0.4
	0615	-0.4		1154	3.7
	1234	4.6		1746	0.2
	1847	-0.6			
4 SA	0111	4.6	19 SU	0029	3.7
	0724	-0.1		0613	0.6
	1329	4.3		1228	3.5
	1951	-0.4		1818	0.3
5 SU	0207	4.6	20 M	0108	3.8
	0835	0.0		0727	0.7
	1428	4.0		1311	3.3
	2053	-0.3		1917	0.4
6 M	0307	4.5	21 TU	0156	3.9
	0939	-0.1		0856	0.7
	1530	3.7		1404	3.3
	2153	-0.3		2050	0.4
7 TU	0407	4.5	22 W	0255	4.0
	1038	-0.2		1001	0.4
	1635	3.6		1515	3.2
	2247	-0.3		2157	0.2
8 W	0508	4.6	23 TH	0404	4.2
	1131	-0.3		1057	0.1
	1735	3.7		1637	3.4
	2339	-0.3		2256	0.0
9 TH	0603	4.7	24 F	0510	4.5
	1222	-0.4		1150	-0.2
	1828	3.8		1742	3.7
				2352	-0.3
10 F	0028	-0.4	25 SA	0609	4.9
	0649	4.6		1242	-0.6
	1310	-0.5		1838	4.0
	1915	3.9			
11 SA	0117	-0.4	26 SU	0048	-0.6
	0733	4.8		0701	5.2
	1354	-0.6		1330	-1.0
	1959	3.9		1929	4.4
12 SU	0202	-0.4	27 M	0140	-0.9
	0814	4.8		0749	5.4
	1436	-0.6		1419	-1.3
	2040	3.9		2019	4.7
13 M	0242	-0.4	28 TU	0231	-1.1
	0854	4.7		0840	5.5
	1516	-0.6		1505	-1.4
	2122	3.9		2110	4.9
14 TU	0321	-0.3	29 W	0321	-1.2
	0931	4.5		0931	5.4
	1552	-0.5		1550	-1.5
	2208	3.9		2203	5.0
15 W	0358	-0.1	30 TH	0410	-1.2
	1011	4.3		1024	5.2
	1625	-0.4		1635	-1.3
	2242	3.8		2258	5.0
			31 F	0459	-0.8
				1120	4.9
				1725	-1.0
				2354	4.9

FEBRUARY

DAY	TIME H.M.	HT. FT.	DAY	TIME H.M.	HT. FT.
1 SA	0554	-0.6	16 SU	0503	0.2
	1213	4.6		1111	3.8
	1818	-0.7		1704	0.1
				2338	4.0
2 SU	0047	4.8	17 M	0535	0.3
	0659	-0.2		1149	3.6
	1309	4.1		1733	0.2
	1919	-0.3			
3 M	0142	4.6	18 TU	0020	4.0
	0808	0.0		0621	0.5
	1405	3.8		1231	3.5
	2024	-0.1		1818	0.4
4 TU	0240	4.4	19 W	0108	4.1
	0915	0.1		0759	0.6
	1508	3.6		1339	3.4
	2129	0.0		1956	0.3
5 W	0341	4.3	20 TH	0211	4.1
	1015	0.0		0926	0.5
	1612	3.5		1439	3.3
	2225	0.0		2125	0.4
6 TH	0444	4.3	21 F	0327	4.2
	1110	-0.1		1026	0.2
	1715	3.5		1607	3.5
	2320	0.0		2213	0.1
7 F	0542	4.4	22 SA	0444	4.5
	1200	-0.2		1123	-0.2
	1811	3.7		1718	3.9
				2334	-0.3
8 SA	0009	-0.1	23 SU	0545	4.8
	0629	4.5		1214	-0.6
	1246	-0.3		1816	4.4
	1856	3.9			
9 SU	0057	-0.2	24 M	0030	-0.7
	0713	4.6		0649	5.2
	1329	-0.5		1306	-1.0
	1937	4.0		1909	4.8
10 M	0141	-0.3	25 TU	0123	-1.0
	0752	4.6		0731	5.4
	1410	-0.5		1354	-1.2
	2015	4.1		1959	5.2
11 TU	0222	-0.4	26 W	0215	-1.3
	0829	4.6		0820	5.4
	1449	-0.6		1441	-1.4
	2053	4.2		2048	5.4
12 W	0300	-0.4	27 TH	0305	-1.4
	0906	4.5		0912	5.3
	1523	-0.5		1526	-1.4
	2129	4.2		2139	5.4
13 TH	0335	-0.3	28 F	0352	-1.3
	0941	4.3		1003	5.1
	1555	-0.4		1612	-1.2
	2203	4.1		2232	5.3
14 F	0407	-0.2			
	1011	4.1			
	1622	-0.2			
	2235	4.1			
15 SA	0435	0.0			
	1043	3.9			
	1643	-0.1			
	2307	4.0			

MARCH

DAY	TIME H.M.	HT. FT.	DAY	TIME H.M.	HT. FT.
1 SA	0442	-1.0	16 SU	0416	-0.1
	1058	4.8		1013	4.0
	1658	-0.9		1613	0.0
	2325	5.1		2223	4.4
2 SU	0533	-0.6	17 M	0443	0.0
	1152	4.4		1043	3.8
	1749	-0.5		1637	0.2
				2259	4.4
3 M	0021	4.9	18 TU	0517	0.2
	0631	-0.2		1124	3.7
	1247	4.1		1708	0.3
	1847	0.0			
4 TU	0115	4.6	19 W	0602	0.4
	0739	0.1		1218	3.6
	1344	3.8		1752	0.5
	1956	0.3			
5 W	0212	4.3	20 TH	0042	4.3
	0848	0.3		0724	0.5
	1443	3.6		1318	3.6
	2103	0.4		1906	0.6
6 TH	0312	4.2	21 F	0133	4.3
	0949	0.3		0853	0.4
	1545	3.5		1429	3.6
	2204	0.4		2106	0.5
7 F	0415	4.1	22 SA	0300	4.3
	1041	0.2		0959	0.2
	1649	3.6		1548	3.9
	2258	0.3		2217	0.2
8 SA	0513	4.2	23 SU	0417	4.5
	1131	0.0		1086	-0.2
	1744	3.8		1658	4.3
	2347	0.1		2315	-0.2
9 SU	0603	4.3	24 M	0524	4.8
	1216	-0.1		1147	-0.5
	1830	4.1		1755	4.8
10 M	0033	0.0	25 TU	0020	-0.5
	0645	4.4		0620	5.0
	1259	-0.3		1238	-0.8
	1848	4.3		1848	5.2
11 TU	0117	-0.2	26 W	0107	-1.0
	0726	4.4		0712	5.2
	1341	-0.4		1328	-1.1
	1946	4.4		1938	5.6
12 W	0158	-0.3	27 TH	0157	-1.2
	0807	4.4		0802	5.3
	1418	-0.4		1416	-1.2
	2022	4.5		2025	5.7
13 TH	0237	-0.4	28 F	0247	-1.3
	0837	4.5		0852	5.2
	1453	-0.4		1502	-1.1
	2054	4.5		2114	5.7
14 F	0313	-0.4	29 SA	0334	-1.2
	0911	4.3		0942	4.9
	1525	-0.3		1547	-0.9
	2126	4.5		2200	5.5
15 SA	0346	-0.3	30 SU	0422	-0.9
	0941	4.2		1031	4.5
	1551	-0.1		1632	-0.6
	2152	4.5		2259	5.3
			31 M	0512	-0.6
				1121	4.2
				1721	-0.2
				2352	4.9

TIME MERIDIAN 75° W. 0000 IS MIDNIGHT. 1200 IS NOON.
HEIGHTS ARE RECKONED FROM THE DATUM OF SOUNDINGS ON CHARTS OF THE LOCALITY WHICH IS MEAN LOW WATER.

Excerpt from tide tables

The National Ocean Survey tide tables come in a series of volumes covering various parts of the U.S. Coast and the rest of the world. Each volume is divided into six main tables:

Table 1 provides full details of times and heights of the tide for HW and LW each day for a specified *reference station.*

Table 2 lists the tidal differences in time and height of the tide to be applied to a specified reference station in order to find the time and height of HW and LW at *subordinate stations.*

Table 3 contains information for finding the height of the tide anywhere between HW and LW.

Table 4 lists the times of sunrise and sunset for different latitudes.

Table 5 provides the means to convert local time into zone or standard time.

Table 6 contains the zone time of moonrise and moonset at specified locations.

TABLE 2.—TIDAL DIFFERENCES AND OTHER CONSTANTS

No.	PLACE	POSITION Lat.	POSITION Long.	Time High water	Time Low water	Height High water	Height Low water	Mean	Spring	Mean Tide Level
		° ′	° ′	h. m.	h. m.	feet	feet	feet	feet	feet
	NEW YORK and NEW JERSEY — Continued									
	Hudson River‡									
	Time meridian, 75°W.			on NEW YORK, p.56						
1513	Jersey City, Pa. RR. Ferry, N. J——	40 43	74 02	+0 07	+0 07	-0.1	0.0	4.4	5.3	2.2
1515	New York, Desbrosses Street——	40 43	74 01	+0 10	+0 10	-0.1	0.0	4.4	5.3	2.2
1517	New York, Chelsea Docks——	40 45	74 01	+0 17	+0 16	-0.2	0.0	4.3	5.2	2.1
1519	Hoboken, Castle Point, N. J——	40 45	74 01	+0 17	+0 16	-0.2	0.0	4.3	5.2	2.1
1521	Weehawken, Days Point, N. J——	40 46	74 01	+0 24	+0 23	-0.3	0.0	4.2	5.0	2.1
1523	New York, Union Stock Yards——	40 47	74 00	+0 27	+0 26	-0.3	0.0	4.2	5.0	2.1
1525	New York, 130th Street——	40 49	73 58	+0 57	+0 55	-0.5	0.0	4.0	4.8	2.0
1527	George Washington Bridge——	40 51	73 57	+0 46	+0 43	-0.6	0.0	3.9	4.6	1.9
1529	Spuyten Duyvil, West of RR. bridge——	40 53	73 56	+0 58	+0 53	-0.7	0.0	3.8	4.5	1.9
1531	Yonkers——	40 56	73 54	+1 09	+1 10	-0.8	0.0	3.7	4.4	1.8
1533	Dobbs Ferry——	41 01	73 53	+1 29	+1 40	-1.1	0.0	3.4	4.0	1.7
1535	Terrytown——	41 05	73 52	+1 45	+1 54	-1.3	0.0	3.2	3.7	1.6
1537	Ossining——	41 10	73 52	+1 55	+2 14	-1.4	0.0	3.1	3.6	1.5
1539	Haverstraw——	41 12	73 58	+1 59	+2 25	-1.6	0.0	2.9	3.4	1.4
1541	Peekskill——	41 17	73 57	+2 24	+3 00	-1.3	+0.3	2.9	3.4	1.7
1543	West Point——	41 24	73 57	+3 16	+3 37	-1.5	+0.3	2.7	3.1	1.6
1545	Newburgh——	41 30	74 00	+3 42	+4 00	-1.5	+0.2	2.8	3.2	1.6
1547	New Hamburg——	41 35	73 57	+4 00	+4 25	-1.5	+0.1	2.9	3.3	1.5
1549	Poughkeepsie——	41 42	73 57	+4 30	+4 43	-1.3	+0.1	3.1	3.5	1.6
1551	Hyde Park——	41 47	73 57	+4 56	+5 09	-1.3	0.0	3.2	3.6	1.6
1553	Kingston Point——	41 56	73 58	+5 16	+5 31	-0.9	-0.1	3.7	4.2	1.7
1555	Tivoli——	42 04	73 56	+5 46	+6 01	-0.8	-0.2	3.9	4.4	1.7
1557	Catskill——	42 13	73 51	+6 37	+6 55	-0.7	-0.3	4.1	4.6	1.7
1559	Hudson——	42 15	73 48	+6 54	+7 09	-0.9	-0.4	4.0	4.4	1.6
				on ALBANY, p.60						
1561	Coxsackie——	42 21	73 48	-1 01	-1 38	-0.5	+0.2	3.9	4.3	2.1
1563	New Baltimore——	42 27	73 47	-0 34	-0 56	-0.1	+0.4	4.1	4.5	2.4
1565	Castleton-on-Hudson——	42 32	73 45	-0 17	-0 29	-0.3	+0.1	4.3	4.7	2.2
1567	ALBANY——	42 39	73 45		Daily predictions			4.6	5.0	2.5
1569	Troy——	42 44	73 42	+0 06	+0 10	+0.1	0.0	4.7	5.1	2.3
	The Kills and Newark Bay			on NEW YORK, p.56						
	Kill Van Kull									
1571	Constable Hook——	40 39	74 05	-0 34	-0 21	0.0	0.0	4.5	5.4	2.2
1573	New Brighton——	40 39	74 05	-0 12	-0 18	0.0	0.0	4.5	5.4	2.2
1575	Port Richmond——	40 38	74 08	+0 05	+0 05	0.0	0.0	4.5	5.4	2.2
1577	Bergen Point——	40 39	74 08	+0 03	+0 03	+0.1	0.0	4.6	5.5	2.3
1579	Shooters Island——	40 39	74 10	+0 06	+0 18	+0.1	0.0	4.6	5.5	2.3
1581	Port Newark Terminal——	40 41	74 08	-0 01	+0 18	+0.6	0.0	5.1	6.1	2.5
1583	Newark, Passaic River——	40 44	74 10	+0 22	+0 52	+0.6	0.0	5.1	6.1	2.5
1585	Passaic, Gregory Ave. bridge——	40 51	74 07	+0 49	+1 57	+0.6	0.0	5.1	6.1	2.5
	Hackensack River									
1586	Kearny Point——	40 44	74 06	+0 09	+0 33	+0.5	0.0	5.0	6.0	2.5
1587	Secaucus——	40 48	74 04	+1 13	+1 09	+0.6	0.0	5.1	6.1	2.6
1588	Little Ferry——	40 51	74 02	+1 22	+1 14	+0.8	0.0	5.3	6.4	2.7
1589	Hackensack——	40 53	74 02	+1 33	+1 55	+0.8	0.0	5.3	6.4	2.6
				on SANDY HOOK, p.64						
	Arthur Kill									
1591	Elizabethport——	40 39	74 11	+0 25	+0 39	+0.3	0.0	4.9	5.9	2.4
1593	Chelsea——	40 36	74 12	+0 34	+0 35	+0.4	0.0	5.0	6.0	2.5
1595	Carteret——	40 35	74 13	+0 23	+0 31	+0.5	0.0	5.1	6.2	2.6
1597	Rossville——	40 33	74 13	+0 17	+0 25	+0.7	0.0	5.3	6.4	2.6
1599	Tottenville——	40 31	74 15	+0 03	+0 13	+0.7	0.0	5.3	6.4	2.6
1601	Perth Amboy——	40 30	74 16	+0 13	+0 19	+0.6	0.0	5.2	6.3	2.6

‡Values for the Hudson River above the George Washington Bridge are based upon averages for the six months May to October, when the fresh-water discharge is a minimum.

Excerpt from tide tables

Chapter 9 **Harbor Pilotage**

Buoyage systems are used in all navigable waters to assist navigators in keeping their craft within safe channels, and to indicate hazards and dangers. There are two major buoyage systems used in U.S. waters; the *United States System*, better known of old as the 'lateral system' and the *Modified U.S. Aids to Navigation System*, which is a part adaptation of the world wide *IALA* system gradually being introduced by all maritime nations.

In all systems, colors and shapes are used to indicate safe or dangerous waters. *Port hand* marks must be passed on the left side of the vessel and *starboard hand* marks must be passed down the right side of the vessel. All reference to port or starboard, left or right are to be taken with the vessel heading in from seaward.

The United States system
Port hand marks are painted black and are can or spar shaped. If numbered they carry odd numbers, commencing at the seaward end and if lighted they display a white or green flashing or occulting light.

Starboard hand marks are painted red and are cone or spar shaped, often referred to as 'nun' buoys. If numbered they carry even numbers, commencing at the seaward end, and if lighted they display a red or white flashing or occulting light.

Other marks are used to indicate middle ground, mid-channel or wreck hazards and are shaped and colored as indicated in the accompanying diagram.

The Modified U.S. Aid system
Port hand marks are painted green and are can shaped. If numbered they carry odd numbers, and if lighted display a green light. If a daymark is fitted it is in the form of a green rectangle.

Starboard hand marks are painted red and are nun shaped. If num-

IALA BUOYAGE SYSTEM 'B'

LATERAL MARKS

PORT HAND

STARBOARD HAND

CARDINAL MARKS

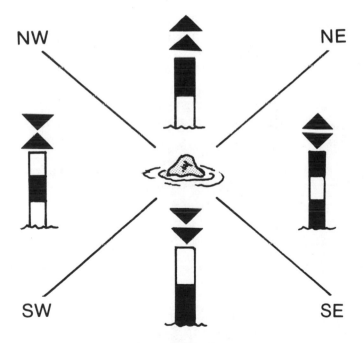

NW

NE

SW

SE

bered they carry even numbers, and if lighted display a red light. If a daymark is fitted it is in the form of a red triangle.

Safe water marks are painted with red and white vertical stripes and are spherical in shape. They do not carry numbers, and if lighted display a white light flashing in a morse code characteristic. If fitted with a topmark it is in the form of a sphere with red and white vertical stripes.

The IALA-B system

Gradually being brought into use in the waters around the U.S. coast, this system is quite different in some ways to the IALA-A system used in Europe and most other parts of the world. The main differences lie in the lateral system where the colors of the A and B systems are reversed.

Lateral marks

These marks usually indicate the limits of a channel or the proximity of shoal water or some other hazard.

Port hand marks are painted green and are in the form of a can, pillar or spar. If lighted they display green lights with any characteristic other than those used for preferred channels.

Starboard hand marks are painted red and are in the form of a cone, pillar or spar. If lighted they display red lights with any characteristic other than those used for preferred channels.

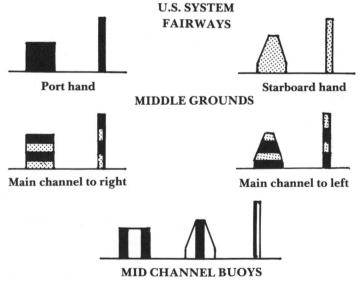

U.S. SYSTEM
FAIRWAYS

Port hand   Starboard hand

MIDDLE GROUNDS

Main channel to right   Main channel to left

MID CHANNEL BUOYS

Preferred channel to starboard marks are painted green with a red horizontal band and are shaped the same as port hand marks. If lighted they display a green light with the characteristic 2+1 (alternating groups of two flashes with a single flash).

Preferred channel to port marks are painted red with a green horizontal band and are shaped the same as starboard hand marks. If lighted they display a red light with the characteristic 2+1.

Cardinal marks

These marks indicate the location of a hazard. They are placed in a cardinal or compass zone relative to the hazard. The marks are usually in the form of a pillar or spar buoy and use colors, topmarks and lights as their means of identification:

North quadrant marks are yellow with a black top and carry two black triangles, one vertically above the other, as topmarks. At night they display a white quick-flashing light.

South quadrant marks are black with a yellow top and carry the same topmark but with both triangles inverted. At night they display a white light quick-flashing in groups of six.

East quadrant marks are black with a horizontal yellow band and carry two black triangles as topmarks, the lower triangle inverted. At night they display a white light quick-flashing in groups of three.

West quadrant marks are yellow with a horizontal black band and carry two black triangles as topmarks, the upper triangle inverted. At night they display a white light flashing in groups of nine.

Other IALA marks

Isolated danger marks are usually pillar or spar shaped and painted with horizontal red and black bands. They carry a topmark of two black spheres in a vertical line and if lighted show a white light flashing in groups of two.

Safe water marks are mostly spherical in shape with vertical red and white bands. The light, if carried, may be Isophase or Occulting, or flash the morse characteristic 'A'.

Special marks, used for a variety of purposes, often non-navigational, such as marking the limits of races, are yellow with a yellow 'X' as a topmark. If fitted with a light it will be yellow with any characteristic that cannot be confused with those of other lights.

IALA east cardinal mark   **IALA special mark**

Other buoys

Different colors, shapes and topmarks are used with buoys to indicate special dangers or areas. These are detailed, as is all information about buoyage systems, in *Sailing Directions* and *Coast Pilot*.

Leaving harbor

It follows that since the colors, shapes and topmarks for buoys are related to the side of the boat on which they must be passed when *entering* harbor from seaward, these must be passed on the *opposite* side of the boat when leaving harbor for sea.

Harbor charts

A large scale harbor chart is essential before entering any port. These may be drawn as full charts (in big commercial harbors there may be a number of charts each covering one harbor area) or as part of small-craft charts, issued by the National Ocean Survey for inland waterways.

Coast Pilots

As with the chart, consultations with the relevant *U.S. Coast Pilot* is vital when entering a strange harbor. Information relating to all buoys, marks and channels is contained in this volume plus other vital details such as tidal movements, port signals, bridge heights and prohibited areas.

96

Dredge signals

Most commercial ports have dredges working somewhere and since these vessels almost invariably work in the channels it is important to be able to identify them and the signals indicating which side they may be passed. There is no system of marking which is adopted internationally, so the *Coast Pilots* or any local harbor publication must be studied to ensure that the navigator will identify a dredge (particularly at night) and know on which side it must be passed.

Port closed signals

Many ports, particularly smaller harbors and estuaries subject to bar conditions, may be closed to entering traffic if conditions necessitate this. Once again, the *Coast Pilots* will provide full details of the signals shown and where they are displayed when the port is closed. This information will not be found on the chart.

Weather signals

Like port closed signals, weather signals may be displayed from a prominent position in a coastal harbor. Details of these signals and what they indicate will be found in the *Coast Pilots*.

Range marks

These are described in detail in Chapter 4. By day, the usual system consists of two triangular marks which must be lined up in

97

order to enter the main channel. They are often painted with bright flourescent colors so that they can be seen even in poor visibility, and are replaced by lights at night. In small ports where a bar may be encountered, the marks may be moved as the channel changes so that they always represent the deepest or best channel for entry. Full details of these marks will be found in the *U.S. Coast Pilot*.

Piles and beacons

Smaller ports, non-commercial harbors and upper reaches of rivers may use piles or beacons instead of buoys to mark channels and hazards. Beacons may be fitted with a colored *daymark* for use in daylight and with a colored light at night. The color code of the light and the shape of the day mark usually follows the buoyage system but once again this should be checked with the chart and Coast Pilot. The usual system is as follows:

Green light or green daymark—pass on the port side of the boat.

Red light or red daymark—pass on the starboard side of the boat.

Because of the numerous waterways around the U.S. coastline and the many different inland waterways, systems may vary and lights and daymarks differ from place to place.

Signal stations

Most harbors of any note have a signal station near the entrance which can be called by signal lamp and asked for information. The location of these signal stations and any special features about times of operating can be found in the *U.S. Coast Pilots*.

Harbors and estuaries can be fraught with dangers for the inexperienced navigator

Navigation beacons come in all shapes and sizes

The chart is the navigator's greatest assest in unfamiliar waters

Channels through shoal waters require careful navigation and correct interpretation of the beacons and lights.

Anchorages

Suitable anchorages and sheltered spots are indicated in the *Coast Pilot* and sometimes on the chart. Along the coastline their position is usually determined by the shelter they afford from the prevailing bad weather winds. In harbors they indicate the best location for both shelter and convenience.

Submarine cables

Most harbors have submarine cables running across the sea bed. Anchorage is strictly prohibited near these for obvious reasons, and they are clearly marked, usually with a large sign on the shore, and with a purple symbol on the chart.

Overhead cables

These can also create hazards for tall-masted yachts proceeding up a river or estuary. They are indicated on the chart and in the *Coast Pilot.*

Commercial shipping

Because they must stick rigidly to deep-water channels, large freighters and other commercial or naval ships have right of way over yachts and pleasure craft within the limits of most major harbors. If there is sufficient water outside the channel, small craft should keep just clear of the marker buoys when encountering such traffic. Failing this, they must hug the *starboard* side of the channel until past and clear. It goes without saying that anchorage is prohibited in such channels.

Yacht races

Entering a popular waterway on a summer Saturday afternoon is somewhat akin to walking into a forest. Yacht races cover much of the water, and while the normal international rules for preventing collision still prevail, it is common courtesy to avoid interfering with races unless absolutely necessary.

Special rules

Every port, especially big commercial ports, has its own special rules concerning not only navigation, but the question of right of way. In many busy ports important vessels such as ferries and hydrofoils have a special right of way over all pleasure craft. Such details can again be found in the *Coast Pilot* or local harbor rule book.

Speed limits

Almost every harbor has speed limits, and while these apply only to fast power boats, their location should be known. Details of speed limits, as well as all navigational details, are contained in the *Coast Pilot* or local harbor rule book.

Crossing a bar

Many coastal harbors and river entrances have a bar across them, and since these can change quickly in both depth and the location of the channel, they present a formidable navigational hazard. Unfortunately, no two bars are the same, and information relating to each must be obtained from local authorities or the *Coast Pilot* in order to ensure safe navigation when entering.

It goes without saying that the condition of a bar deteriorates rapidly with big seas or bad weather conditions. The breaking waves on a bar are difficilt to see when approached from seaward, and many a seaworthy craft has been lost by attempting to enter what appeared to be a relatively safe estuary.

The *U.S. Coast Pilots* carry information relating to such harbor entrances, but if in doubt a radio call to the U.S. Coastguard or the local port authorities will provide the necessary guidance to make a safe entry.

Chapter 10   Electronic Navigation

There are many uses for electronic navigation in coastal waters. Apart from the more obvious uses, such as navigating through fog and other 'blind' conditions, there are times when navigation by electronics is easier and often more accurate than by visual means.

A typical example of this would be when sailing well offshore along a bleak, low coastline where prominent objects are few and far between. Visual fixes may be possible, but would not be accurate, whereas a fix from one of a number of electronic aids could pinpoint the boat's position with reasonable accuracy.

The most widely used of these electronic aids for navigating along a coastline are:

Radio Direction Finder (RDF)
Radar
Satnav (Satellite Position Finding)
Omega
Loran C
The depth sounder
Decca

Radar antenna can be somewhat cumbersome when fitted to a sailing yacht

There are limitations with all these electronic aids and not all are available or useable on every coastline. However, one or the other are usually available and often a combination of electronic and visual navigation systems may be used to obtain accurate fixes.

Radio Direction Finding

Oldest of the electronic family of navigational aids, radio direction finding has now become a sophisticated and accurate method of position finding for craft in coastal waters as well as for boats some distance out to sea. Radio beacons are established along most coastlines with a range well out to sea and are often linked into a chain to provide total coverage along each coastal strip.

The tuneable antenna of an RDF receiver reads off against a compass card when the 'null' signal is obtained, this giving the bearing of the beacon

The radio direction finder (RDF) is a medium frequency radio receiver fitted with a directional antenna. When a signal is picked up from a shore radiobeacon, the antenna is rotated and the signal tuned until a bearing of the beacon from the boat can be taken. Since the position of all transmitting beacons are marked on the chart, a series of bearings of different beacons can be plotted to obtain a cross bearing fix of the boat's position.

To facilitate use of the RDF, radiobeacons are usually grouped together and transmit on the same frequency one after the other in sequence. This avoids overlap of the signals and saves constant retuning from station to station. Some receivers operate automatically, making the task even easier. Since radiobeacons usually

103

operate around the clock in all weathers, they are a very useful aid to navigation and within a range of 100 miles or so should provide bearings accurate to within a couple of degrees.

The transmitted signal is identified by its characteristic which is in the form of a Morse code signal in dots and dashes. The station can be identified from D M A H T C Publications Nos. 117A and 117B, *Radio Navigational Aids,* and its geographical location is marked on the chart by a purple circle and the letters RDF.

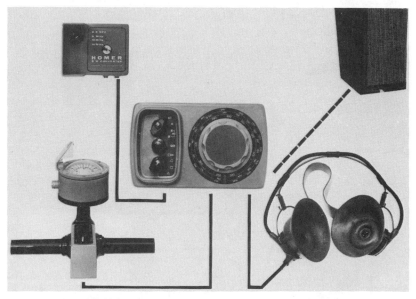

A sophisticated RDF system for use in a yacht or small craft

The procedure for obtaining a fix of the boat's position by RDF is as follows:

1 Tune in the receiver and identify the beacon from its morse symbol.

2 Tune the antenna until the signal fades out completely. This is known as the 'null'.

3 Read off the bearing of the beacon. This may be done from a magnetic compass fitted to the receiver or from a pelorus which indicates the relative bearing of the beacon to the boat's head.

4 Convert this compass bearing to true and lay off on the chart.

5 Repeat the procedure with other beacons in the sequence. The result will be a cross bearing fix of the boat's position.

Radar

As described in Chapter 2, radar can be of great use in navigating small craft, particularly at night on a dark coastline, or when weather is bad and shore objects are not visible. However, radar does have its limitations and it is important to keep this in mind.

A typical small craft PPI radar display unit

The pulse transmitted by the radar 'echoes' back off the shoreline and other objects within range and displays them on a plan position indicator (PPI) screen. This presents the region surrounding the boat as a map with the boat in the center and all objects within range located in their relative positions.

One of the limitations of the equipment is that different objects give different echoes and the PPI does not always provide an accurate picture of the situation. A cliffy section of the coastline will show up clearly on the radar screen, but a low, mangrove area will give only a faint echo, perhaps none at all. Similarly, radar cannot see round corners, and a low headland located behind a more prominent headland will be 'blanketed' from the radar pulse and will not appear on the screen.

A coastline with such characteristics will therefore be considerably distorted when reproduced on the radar screen. Careful interpretation of the PPI display is necessary before accepting the picture as being accurate. There are other areas where radar is adversely affected and before using such equipment in practice, a wise navigator will take a radar course or obtain advice in interpreting the PPI screen.

The radar antenna is usually housed in a 'pod' or 'transponder'

Despite these limitations, the radar set is a very useful piece of navigational equipment. For position finding it can be used in a number of ways, the most common of which are as follows:

Fix by radar bearing and range.

1 Identify a prominent object on the radar screen and check its location on the chart. Accurate identification is essential.

2 Move the pelorus ring around the screen until the radial cursor line cuts the echo of the required object. Read off the relative bearing. Apply the course steered to convert it to a compass bearing

3 Obtain the distance of the object from the boat by means of the range rings or the variable range marker (VRM).

4 Lay off the bearing of the object on the chart and measure along it the range obtained from the radar screen. The result is the fix of the boat's position

Note: If the object is visible to the eye, a more accurate bearing will be obtained by using the hand-bearing compass instead of the radar cursor. A visual bearing together with a radar range provides a very accurate fix.

Fix by radar ranges.

This method requires identification of two or more prominent objects preferably spaced around 90° apart.

1 Identify the objects on the radar screen and also the chart.

Satellite navigation receivers are ideally suited for use on small craft

2 Using the VRM or the range rings, obtain a distance off each object on the radar screen.

3 Using a pair of compasses, draw an arc of a circle, the same radius as the radar range, around each object.

4 The intersection of the arcs is the fix of the boat's position.

There are many other uses for radar in navigation, but there is no room in a book such as this to describe them in detail. The two plotting methods mentioned are the most widely used aboard small craft for coastal navigation.

Satellite navigation

Although generally accepted as a navigational aid for ocean crossing, Satnav provides the coastal navigator with yet another string to his electronic bow. This relatively new method of position finding is extremely accurate and can be used in virtually any waters, inshore or offshore.

Satnav in its most basic form, is a receiver which obtains information from an orbiting navigational satellite and uses this information to compute the boat's position. The latest developments of this system, known as Global Position System (GPS) claim an accuracy anywhere in the world to within a few metres. This system is due to become fully operational in 1988, in the meantime the present Navsat system provides an accuracy claimed at 0.1 nautical mile.

Although there are a number of satellites available for navigation, only one is required for position finding at any given time. As the satellite sweeps up over the horizon and passes across the sky, the frequency of its transmissions changes in just the same way that the pitch of an aircraft's engine changes audibly as it passes over an observer on the ground. By identifying the satellite's frequency 'pitch', and knowing the satellite's position in its orbit at that time, the receiver can compute the location of the boat on the earth's surface.

Accurate satellite navigation is subject to only a few limitations, one of which is that a fix can only be obtained if a satellite is above the horizon. The receiver must be prepared with a DR position fed in or updated at the time of a fix. The height of the antenna must be taken into consideration when tuning the set. Most accurate readings are obtained when the satellite is at an altitude of between 15° and 75°.

With these factors in hand, it requires only the press of a button for the instrument to produce an extremely accurate latitude and longitude readout.

Omega

This is a modern and sophisticated radio navigation system which is more commonly used for ocean crossing than for coastal passages. It provides world-wide coverage from eight transmitters located around the globe using a phase-difference technique with very low frequency transmissions. Lattice charts are used for plotting the readings taken from the Omega receiver which is accepted as having an accuracy of about one nautical mile at any point on the globe.

Loran C

Another radio navigation aid, this time with a shorter range, which is of particular use in coastal waters. Loran C is extremely accurate and easy to use. Most receivers provide a straight latitude and longitude readout which can be transferred directly to a coastal chart.

Depth sounder

Obtaining a fix of the boat's position using nothing but the depth sounder may be difficult unless the contours of the sea bed are very marked and provide steep contrasts. More often than not the sounder is used in conjunction with another instrument, such as a compass, to find the boat's position.

A typical case would be a combined bearing and sounding. In

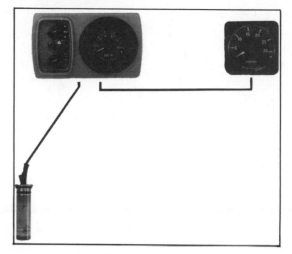

Illustrating a typical depth sounding system

this fix a sounding of the water is obtained at the same time as a compass bearing is taken off a prominent object. When the compass bearing is laid off on the chart, the boat's position can be determined by the location of the sounder reading (converted for tide height) along the bearing.

Using only the depth sounder, a method known as *line of soundings* is used. The procedure, which requires frequent distinct changes in soundings, is as follows:

1 At intervals during the boat's progress along her courseline, note any distinct changes in the depth sounding and note also the log reading at that time.

2 Adjust the soundings for tide height and mark them along the edge of a piece of paper at intervals equal to the log distance run between readings.

3 Place the paper on the chart and align the edge with the boat's true course. Slide it across the chart until the depth readings on the paper coincide with identical readings on the chart.

4 Draw a line along the edge of the paper. This will represent the boat's track over the period in which the soundings were taken.

Decca

One of the finest electronic coastal navigation systems, Decca in North America is unfortunately limited to areas of the north

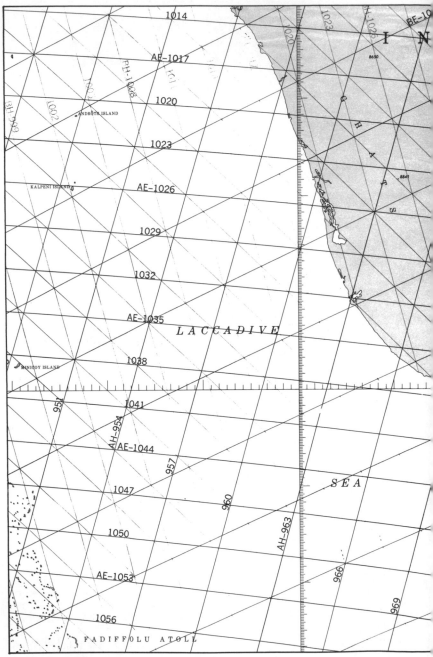

An Omega chart

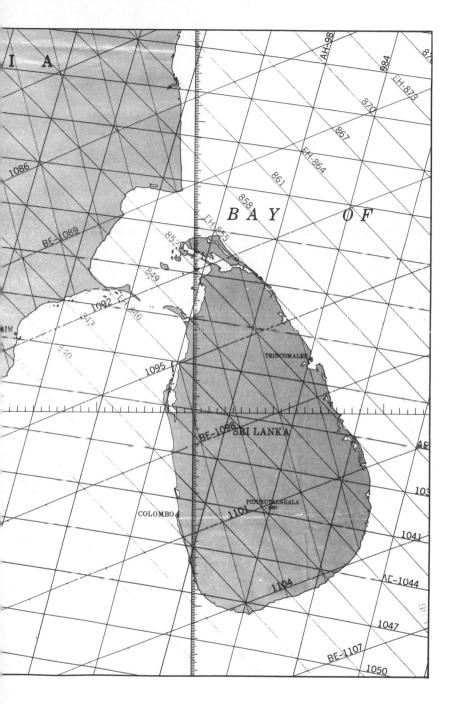

BAY OF

TRINCOMALEE

SRI LANKA

COLOMBO

PIDURUTALAGALA

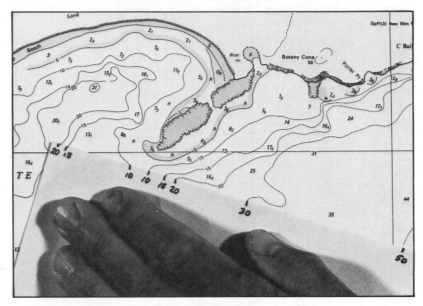

Depths taken from the sounder, when compared with those on the chart, indicate the boat's track

eastern United States and eastern Canada. It is a popular system on the other side of the Atlantic, covering most European waters.

Decca provides a series of readings from automatic receivers. These readings are transferred to a lattice chart and provide a very accurate position. The Decca Yacht Navigator, designed for use in small-craft, provides a latitude and longitude readout similar to Loran C.